To

Renate

and

Larissa

Günter Ludyk

Stability of Time-Variant Discrete-Time Systems

Advances in Control Systems and Signal Processing

Editor: Irmfried Hartmann

Volume 5

Günter Ludyk

Stability of Time-Variant Discrete-Time Systems

With 20 Figures

Springer Fachmedien Wiesbaden GmbH

CIP-Kurztitelaufnahme der Deutschen Bibliothek

Ludyk, Günther:
Stability of time-variant discrete-time systems /
Günter Ludyk.

(Advances in control systems and signal
processing; Vol. 5)
ISBN 978-3-528-08911-5   ISBN 978-3-663-13932-4 (eBook)
DOI 10.1007/978-3-663-13932-4

NE: GT

Editor:

Dr.-Ing. I. Hartmann
Prof. für Regelungstechnik und Systemdynamik
Technische Universität Berlin
Hardenbergstraße 29c
1000 Berlin 12, West Germany

Author:

Dr. Ing. G. Ludyk
o. Prof. für Regelungstheorie
Universität Bremen
Postfach 330440
2800 Bremen, West Germany

ISSN 0724-9993

ISBN 978-3-528-08911-5

Preface

In this monograph some stability properties of linear,
time-variant, discrete-time systems are summarized, where
some properties are well known, some are little-known facts,
and a few may be new. Models for this treatise on the asymp-
totical behaviour of solutions of difference equations are
the commonly known excellent books of CESARI [3] and CONTI [5].

In the tables of Chapter 1 the definitions and the essen-
tial statements on stability of discrete-time systems are
summarized, such that Chapter 2 to 5 may be regarded as
explaining appendices for these tables.

I am grateful to Paul Ludyk, who typed and corrected the
manuscript with great skill and patience, and Alois Ludyk,
who drew the figures with great artistic skill.

Bremen, January 1985   Günter Ludyk

Contents

Notations

$A(k)$, A_o, $A_1(k)$, $A_2(k)$ system matrix

$a_{ij}(k)$ element of $A(k)$

$B(k)$ input matrix

$C(k)$ output matrix

c constant

$\det$ determinant

$\dim$ dimension

$\mathcal{E}\mathcal{S}$ see Def.3-50

e_i basis vector

$F(t)$ system matrix of a continious-time system

$f(k)$ forcing vector

$f(.)$ forcing vector sequence

$G(k,i)$ impulse response matrix

$H(k)$ hermitian matrix $A^*(k)A(k)$

I identity matrix

J JORDAN matrix

J_i, J_{ij} submatrix of J

K time interval

$K(k)$ observer matrix

k discrete time

k_o initial time

kernel   null space

$L(k)$ feedback matrix

ℓ_K^p see Def.4-4

$\ell^p\text{-BIBO-}\mathcal{S}$ see Def.5-28

$\ell^p\text{-BIBS-}\mathcal{S}$ see Def.5-28

$\ell^p\text{-BIBO-stable}$ see Def.4-17

$\ell^p\text{-BIBS-stable}$ see Def.4-10

$\ell^P \mathscr{S}$ see Def.3-46

M(k+N,k)   observability gramian
m   dimension of input vector u(k)

N   integer
n   dimension of state vector

R(k,N)   observability matrix
r   dimension of output vector y(k)

$\mathscr{S}$ see Def.3-2
S(k,N)   reachability matrix
$\mathscr{S}$B   see Def.3-14

T(k)   transformation matrix
T(superscript) denotes transpose

$\mathfrak{U}$A$\mathscr{S}$ see Def.3-56
$\mathfrak{U}$Dt   see Def.5-28
$\mathfrak{U}\mathscr{S}$ see Def.3-19
$\mathfrak{U}\mathscr{S}$z   see Def.5-28
u(k)   input vector
u(.)   input vector sequence
$u_{[k_o,N-1]}$ input vector sequence
W(k,k-N)   reachability gramian
W_1, W_2 submatrix of W(k,k-N)
x(k)   state vector
x_o initial state vector
$\hat{x}(k)$ observer state vector
$\tilde{x}(k)$ observer error vector

y(k)   output vector
$\hat{y}(k)$ observer output vector
y(k,N)   output sequence vector

z(k)   transformed state vector

α least border of time interval K
β constant
δ constant
ε constant

$\Phi(k,i)$ state transition matrix

λ eigenvalue

λ_A eigenvalue of A

λ_H eigenvalue of H(k)

ω highest border of time interval K

$\|\cdot\|$ norm

*(superscript)   denotes complex conjugate transpose

$\emptyset$ empty set

$\blacksquare$ the end of a proof, or example

1 Introduction and Summary

A *discrete-time* system is one in which the signals of the
system are defined only at discrete instants of time. Discrete-
time systems arise on the one side from modelling systems,
that are inherently digital such as economic systems, popu-
lation models, radar tracking systems, digital filters, and
computer-controlled systems, where the inputs and outputs
are periodically sampled; and on the other side from approxi-
mation of continuous-time models, where the inputs and outputs
are approximated as piecewise constant functions.

The use of microprocessors and other digital devices to
realize control systems has become very common, and in view
of the rapid development, and decreasing costs of digital
technology, this trend is likely to continue and perhaps
accelerate.

Continuous-time systems are usually described by differential equations, which relate the derivatives of the dynamic
variables to the current values. The dynamic behaviour viewed
in discrete-time is usually described by *difference equations*
relating the values of the variables at one time to the values
at adjacent times. In these cases, the variables can be viewed
as sequences defined on the set $\mathbb{Z}$ of integers.

In this monograph some stability properties of linear,
time-variant, discrete-time dynamical systems are represented;
the table of contents will provide a general idea of the scope
of the monograph. Chapter 2 deals principally with the mathematical description of linear discrete-time systems. After a
brief introduction of state equations some properties of the
state transition matrix and some estimations of the norm of
this matrix are given, which are needed in the following
chapters. Chapter 3 deals with stability of solutions of

state equations for unforced systems. Solutions of state
equations for forced discrete-time systems and their asymptotically behaviour are covered in chapter 4. Chapter 5 explores
the relations between internal and external stability in some
depth. Section 5.1 to 5.4 discusses structural properties as
reachability, observability, stabilizability, and detectability. These properties are needed to explain some relations
between internal and external stability in the last sections
of this final chapter.

For easy reference the stability definitions for unforced
systems are summarized in Table 1 and 2, and for forced
systems in Table 3 and 4. The fundamental theorems on the
stability of unforced systems are given in Table 5, and the
relations between the different stability classes in Table 6.
In Table 7, 8, 9, and 10 resp. necessary respectively sufficient stability conditions are summarized, which were evalu-
ated in Chapter 3 for unforced time-variant, discrete-time,
linear systems. The fundamental theorems on the stability
of forced systems are given in Table 11 and the relations
between internal and external stability derived in Capter 4
and 5, are given in Table 12.

Stability Definitions
for solutions of $x(k+1) = A(k)x(k)$

			Def.
The solution $x(k;k_0,x_0)$ is said to be	*stable*	$:\Leftrightarrow \underset{\varepsilon>0}{\forall}\ \underset{k_0\in K}{\forall}\ \underset{\delta(k_0,\varepsilon)>0}{\exists}\ \underset{k\in[k_0,\omega)}{\forall}\ \|\tilde{x}_0-x_0\|<\delta(k_0,\varepsilon)\Rightarrow\|\tilde{x}(k;k_0,\tilde{x}_0)-x(k;k_0,x_0)\|<\varepsilon$	Def. 3-1
	bounded	$:\Leftrightarrow \underset{k_0\in K}{\forall}\ \underset{\mu(k_0)>0}{\exists}\ \underset{k\in[k_0,\omega)}{\forall}\ \|x(k;k_0,x_0)\|\leq\mu(k_0)$	Def. 3-10
	uniformly stable	$:\Leftrightarrow \underset{\varepsilon>0}{\forall}\ \underset{\delta(\varepsilon)>0}{\exists}\ \underset{k_0\in K}{\forall}\ \underset{k\in[k_0,\omega)}{\forall}\ \|\tilde{x}_0-x_0\|<\delta(\varepsilon)\Rightarrow\|\tilde{x}(k;k_0,\tilde{x}_0)-x(k;k_0,x_0)\|<\varepsilon$	Def. 3-21
	asymptotically stable	$:\Leftrightarrow$ i) $x(k;k_0,x_0)$ is stable and ii) $\underset{k_0\in K}{\forall}\ \underset{k\to\omega}{\lim}\|\tilde{x}(k;k_0,\tilde{x}_0)-x(k;k_0,x_0)\|=0$	Def. 3-33
		$:\Leftrightarrow$ { i) $x(k;k_0,x_0)$ is stable and ii) $\underset{k_0\in K}{\forall}\ \underset{\varepsilon>0}{\forall}\ \underset{\delta(\varepsilon,k_0)>0}{\exists}\ \underset{T(k_0,\varepsilon)\geq0}{\exists}\ \underset{k\in[k_0+T,\omega)}{\forall}\ \|\tilde{x}_0-x_0\|<\delta\Rightarrow\|\tilde{x}(k;k_0,\tilde{x}_0)-x(k;k_0,x_0)\|<\varepsilon$	Def. 3-34
	uniformly asymptotically stable	$:\Leftrightarrow$ i) $x(k;k_0,x_0)$ is unif.stable and ii) $\underset{k_0\in K}{\forall}\ \underset{k\to\omega}{\lim}\|\tilde{x}(k;k_0,\tilde{x}_0)-x(k;k_0,x_0)\|=0$	Def. 3-56
		$:\Leftrightarrow$ { i) $x(k;k_0,x_0)$ is uniformly stable and ii) $\underset{\varepsilon>0}{\forall}\ \underset{\delta(\varepsilon)>0}{\exists}\ \underset{T(\varepsilon)\geq0}{\exists}\ \underset{k_0\in K}{\forall}\ \underset{k\in[k_0+T,\omega)}{\forall}\ \|\tilde{x}_0-x_0\|<\delta\Rightarrow\|\tilde{x}(k;k_0,\tilde{x}_0)-x(k;k_0,x_0)\|<\varepsilon$	Def. 3-57
	exponentially stable	$:\Leftrightarrow \underset{k_0\in(\alpha,\infty)}{\forall}\ \underset{c>0}{\exists}\ \underset{\beta\in[0,1)}{\exists}\ \underset{k\geq k_0}{\forall}\ \|x(k;k_0,x_0)\|\leq c\,\beta^{k-k_0}\|x_0\|$	Def. 3-51
	short time bounded	$:\Leftrightarrow \underset{k_0\in(\alpha,\omega-N)}{\forall}\ \underset{k\in[k_0,k_0+N]}{\forall}\ \|x_0\|<\delta\Rightarrow\|x(k;k_0,x_0)\|<\varepsilon$ for given $\varepsilon>0,\delta>0,N>0$	Def. 3-13

DEFINITION OF SYSTEM CLASSES
for systems $x(k+1) = A(k)x(k)$

Definition	
$A(k) \in S \quad :\Leftrightarrow \quad \underset{k_0 \in K}{\forall} \quad \underset{c(k_0)>0}{\exists} \quad \underset{k \in [k_0,\omega)}{\forall} \ \|\Phi(k,k_0)\| < c(k_0)$	Definition 3-2
$A(k) \in SB \quad :\Leftrightarrow \quad \underset{k_0 \in (\alpha,\omega-N)}{\forall} \quad \underset{k \in [k_0,k_0+N]}{\forall} \ \|\Phi(k,k_0)\| < \varepsilon/\delta \quad \text{for given } \varepsilon>0,\delta>0,N>0$	Definition 3-14
$A(k) \in US \quad :\Leftrightarrow \quad \underset{c>0}{\exists} \quad \underset{k_0 \in K}{\forall} \quad \underset{k \in [k_0,\omega)}{\forall} \ \|\Phi(k,k_0)\| < c$	Definition 3-19
$A(k) \in AS \quad :\Leftrightarrow \quad \text{i) } A(k) \in S \ \text{ and ii) } \underset{k_0 \in K}{\forall} \ \underset{k\to\omega}{\lim} \Phi(k,k_0) = 0$	Definition 3-35
$A(k) \in UAS \quad :\Leftrightarrow \quad \text{i) } A(k) \in US \ \text{ and ii) } \underset{k_0 \in K}{\forall} \ \underset{k\to\omega}{\lim} \Phi(k,k_0) = 0$	Definition 3-56
$A(k) \in ES \quad :\Leftrightarrow \quad \underset{c>0}{\exists} \quad \underset{\beta \in [0,1)}{\exists} \quad \underset{k_0 \in K}{\forall} \quad \underset{k \in [k_0,\omega)}{\forall} \ \|\Phi(k,k_0)\| \leq c \, \beta^{k-k_0}$	Definition 3-50
$A(k) \in \ell^p S \quad :\Leftrightarrow \quad \underset{k_0 \in K}{\forall} \quad \underset{c_p(k_0)>0}{\exists} \quad \underset{k \in [k_0,\omega)}{\forall} \ \left[\sum_{i=k_0}^{k-1}\|\Phi(k,i)\|^p\right]^{1/p} < c_p(k_0) \, , \ p \in [1,\infty]$	Definition 3-46

Table 3

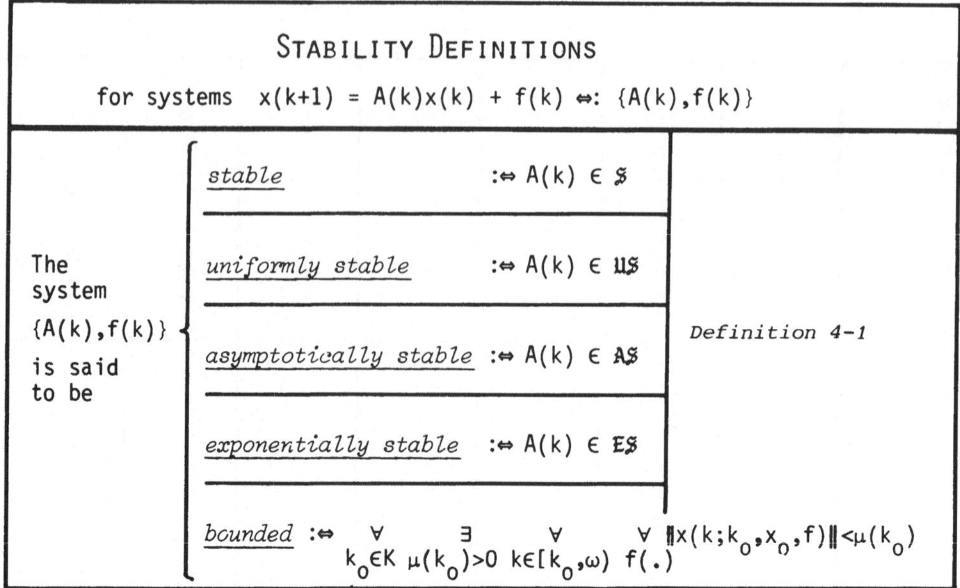

STABILITY DEFINITIONS

for systems $\quad x(k+1) = A(k)x(k) + f(k) \Leftrightarrow: \{A(k),f(k)\}$

The system $\{A(k),f(k)\}$ is said to be		
stable	$:\Leftrightarrow A(k) \in \mathcal{S}$	Definition 4-1
uniformly stable	$:\Leftrightarrow A(k) \in \mathcal{US}$	
asymptotically stable	$:\Leftrightarrow A(k) \in \mathcal{AS}$	
exponentially stable	$:\Leftrightarrow A(k) \in \mathcal{ES}$	
bounded	$:\Leftrightarrow \underset{k_0 \in K}{\forall} \ \underset{\mu(k_0)>0}{\exists} \ \underset{k \in [k_0,\omega)}{\forall} \ \underset{f(.)}{\forall} \ \|x(k;k_0,x_0,f)\| < \mu(k_0)$	

Table 4

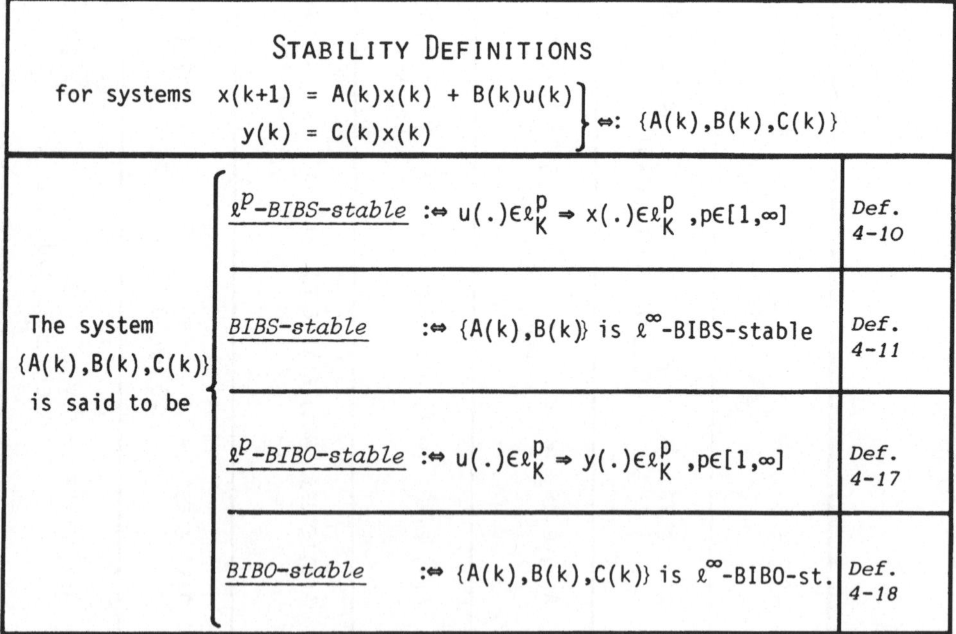

STABILITY DEFINITIONS

for systems $\quad \left.\begin{array}{l} x(k+1) = A(k)x(k) + B(k)u(k) \\ y(k) = C(k)x(k) \end{array}\right\} \Leftrightarrow: \{A(k),B(k),C(k)\}$

The system $\{A(k),B(k),C(k)\}$ is said to be		
ℓ^p-_BIBS-stable_ $\quad :\Leftrightarrow u(.) \in \ell^p_K \Rightarrow x(.) \in \ell^p_K \ , p \in [1,\infty]$		Def. 4-10
BIBS-stable $\quad :\Leftrightarrow \{A(k),B(k)\}$ is ℓ^∞-BIBS-stable		Def. 4-11
ℓ^p-_BIBO-stable_ $\quad :\Leftrightarrow u(.) \in \ell^p_K \Rightarrow y(.) \in \ell^p_K \ , p \in [1,\infty]$		Def. 4-17
BIBO-stable $\quad :\Leftrightarrow \{A(k),B(k),C(k)\}$ is ℓ^∞-BIBO-st.		Def. 4-18

<u>Table 5</u>

<table>
<tr><td colspan="3" align="center">FUNDAMENTAL THEOREMS
for solutions of $x(k+1) = A(k)x(k)$</td></tr>
<tr><td rowspan="6">Every
solution
$x(k;k_0,x_0)$
is</td><td>stable $\Leftrightarrow$ $A(k) \in \mathcal{S}$</td><td>Theorem 3-7</td></tr>
<tr><td>bounded $\Leftrightarrow$ $A(k) \in \mathcal{S}$</td><td>Theorem 3-12</td></tr>
<tr><td>short time bounded $\Leftrightarrow$ $A(k) \in \mathcal{S}\mathbf{B}$</td><td>Theorem 3-15</td></tr>
<tr><td>uniformly stable $\Leftrightarrow$ $A(k) \in \mathcal{US}$</td><td>Theorem 3-23</td></tr>
<tr><td>asymptotically stable $\Leftrightarrow$ $A(k) \in \mathcal{AS}$</td><td>Theorem 3-37</td></tr>
<tr><td>exponentially stable $\Leftrightarrow$ $A(k) \in \mathcal{ES}$</td><td>Theorem 3-52</td></tr>
</table>

<u>Table 6</u>

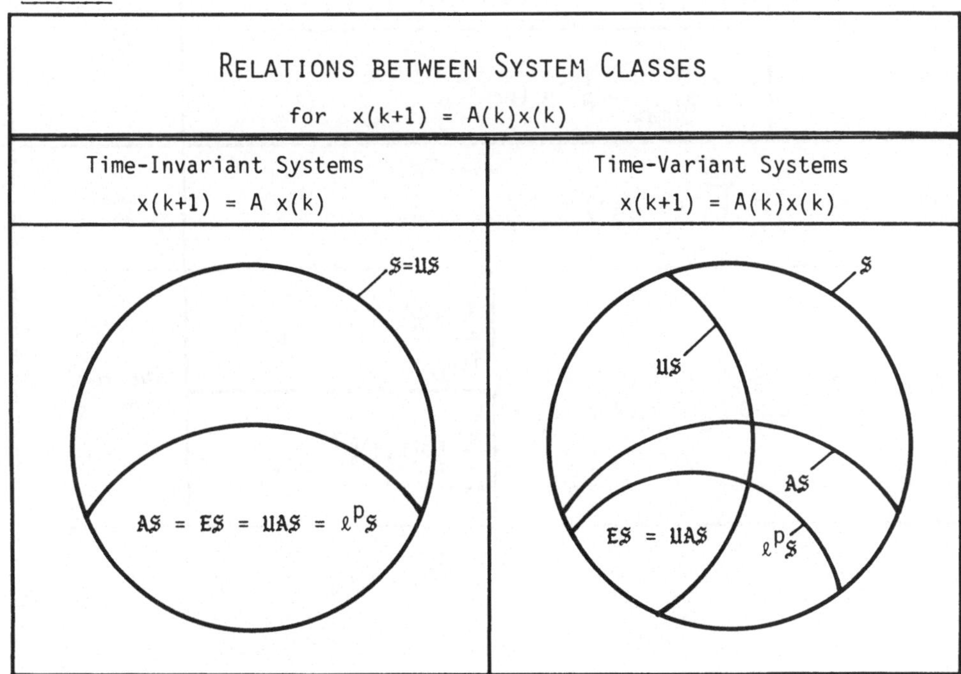

<u>Table 7</u>

NECESSARY STABILITY CONDITIONS

$A(k) \in \mathcal{S} \Rightarrow \begin{cases} \lim\limits_{k\to\omega} \prod\limits_{i=k_0}^{k-1} \lambda_{Hmin}^{1/2}(i) < \infty \\[2ex] \lim\limits_{k\to\omega} \sum\limits_{i=k_0}^{k-1} \frac{1}{2} \ln \lambda_{Hmin}(i) < \infty \end{cases}$		*Lemma 3-8*
$A(k) \in \mathcal{SB} \Rightarrow \forall_{k_0\in(\alpha,\omega-N)} \; \forall_{k\in[k_0,k_0+N]} \begin{cases} \prod\limits_{i=k_0}^{k-1} \lambda_{Hmin}^{1/2}(i) \leq \varepsilon/\delta \\[2ex] \sum\limits_{i=k_0}^{k-1} \ln \lambda_{Hmin}(i) \leq 2\ln\varepsilon/\delta \end{cases}$		*Theorem 3-17*
$A \in \mathcal{SB} \Rightarrow \begin{cases} \left|\lambda_A\right|_{max} \leq (\varepsilon/\delta)^{1/N} \\[2ex] \ln \left|\lambda_A\right|_{max} \leq \frac{1}{N} \ln (\varepsilon/\delta) \end{cases}$		*Lemma 3-18*
$A(k) \in \mathcal{AS} \Rightarrow \lim\limits_{k\to\omega} \prod\limits_{i=k_0}^{k-1} \det A(i) = 0$		*Lemma 3-42*
$A(k) \in \mathcal{ES} \Rightarrow \exists_{c_1>0} \; \exists_{c_2>0} \; \forall_{k_0\in K} \; \forall_{k\in[k_0,\infty)} \begin{cases} \sum\limits_{i=k_0}^{k-1} \|\Phi(k,i)\| < c_1 \\[2ex] \sum\limits_{i=k_0}^{k-1} \|\Phi(k,i)\|^2 < c_2 \end{cases}$		*Theorem 3-54*

<table>
<tr><td colspan="2" align="center">SUFFICIENT STABILITY CONDITIONS I</td></tr>
<tr><td>

$$\forall_{k \geq k_0} \prod_{i=k_0}^{k-1} \lambda_{Hmax}^{1/2}(i) < \infty \;\Rightarrow\; A(k) \in \mathcal{S}$$

$$\forall_{k \geq k_0} \sum_{i=k_0}^{k-1} \ln \lambda_{Hmax}^{1/2}(i) < \infty \;\Rightarrow\; A(k) \in \mathcal{S}$$

</td><td>Lemma
3-8</td></tr>
<tr><td>

$$\left.\begin{array}{l} \text{(i)}\quad A_0 \in \mathcal{S} \\[2em] \text{(ii)}\quad \sum_{i=k_0}^{\omega} \|A_1(i)\| < \infty \end{array}\right\} \;\Rightarrow\; [A_0 + A_1(k)] \in \mathcal{S}$$

</td><td>Theorem
3-29</td></tr>
<tr><td>

$$\left.\begin{array}{l} \text{(i)}\quad A_0(k) \in \mathcal{S} \\[2em] \text{(ii)}\quad \sum_{i=k_0}^{\omega} \|A_1(i)\| < \infty \end{array}\right\} \;\Rightarrow\; [A_0(k) + A_1(k)] \in \mathcal{S}$$

</td><td>Theorem
3-30</td></tr>
<tr><td>

$$\forall_{k_0 \in (\alpha, \omega-N)} \forall_{k \in [k_0, k_0+N]} \prod_{i=k_0}^{k-1} \lambda_{Hmax}^{1/2}(i) \leq \varepsilon/\delta \;\Rightarrow\; A(k) \in \mathcal{SB}$$

$$\forall_{k_0 \in (\alpha, \omega-N)} \forall_{k \in [k_0, k_0+N]} \sum_{i=k_0}^{k-1} \ln \lambda_{Hmax}(i) \leq 2\ln(\varepsilon/\delta) \;\Rightarrow\; A(k) \in \mathcal{SB}$$

</td><td>Theorem
3-16</td></tr>
</table>

SUFFICIENT STABILITY CONDITIONS II

$$\bigvee_{k_0 \in K} \ \bigvee_{k \in [k_0,\omega)} \ \prod_{i=k_0}^{k-1} \lambda_{Hmax}^{1/2}(i) < \infty \ \Rightarrow \ A(k) \in US$$

Lemma 3-20

$$\bigvee_{k_0 \in K} \ \bigvee_{k \in [k_0,\omega)} \ \sum_{i=k_0}^{k-1} \ln \lambda_{Hmax}^{1/2}(i) < \infty \ \Rightarrow A(k) \in US$$

(i) $a_{ij}(k)$ has bounded variation

(ii) $|\lambda_i(k)| \leq 1$

(iii) the eigenvalues of $A_0 := \lim_{k \to \infty} A(k)$,
    whose absolute value is equal
    to one, are simple

$\Bigg\} \ \Rightarrow \ A(k) \in US$

Theorem 3-26

$$\bigvee_{k_0 \in K} \ \underset{k_1 \in [k_0,\omega)}{\exists} \ \bigvee_{k \in [k_1,\omega)} \lambda_{Hmax}(k) \leq 1 \ \Rightarrow \ A(k) \in US$$

Theorem 3-28

$$\bigvee_{k_0 \in K} \ \prod_{i=k_0}^{\omega-1} \lambda_{Hmax}^{1/2}(i) = 0 \ \Rightarrow \ A(k) \in AS$$

Lemma 3-39

$$\bigvee_{k_0 \in K} \ \underset{k \in [k_0,\omega)}{\exists} \ \lambda_{Hmax}(k) = 0 \ \Rightarrow \ A(k) \in AS$$

Lemma 3-40

$$\underset{k_0 \in (\alpha,\omega)}{\bigvee} \ \underset{k_1 \geq k_0}{\exists} \ \underset{k \geq k_1}{\bigvee} \lambda_{Hmax}(k) < 1 \ \Rightarrow \ A(k) \in AS$$

Theorem 3-45

(i) $A(k) \in ES$, i.e. $\underset{c>0}{\exists} \ \underset{\beta \in [0,1)}{\exists} \ \underset{k \geq k_0}{\bigvee}$
    $\|\Phi(k,k_0)\| \leq c \ \beta^{k-k_0},$

(ii) $\|A'(k)\| \leq \delta$ for $k \geq k_0$,

(iii) $(\alpha + \delta c) \in [0,1)$

$\Bigg\} \ \Rightarrow \ [A(k)+A'(k)] \in ES$

Theorem 3-60

Table 10

<table>
<tr>
<td colspan="3" align="center">NECESSARY AND SUFFICIENT
STABILITY CONDITIONS I</td>
</tr>
<tr>
<td rowspan="2">Time-invariant
systems

$x(k+1) = A\,x(k)$</td>
<td>(i) $\forall_{\lambda_i}\ |\lambda_i| \leq 1$

(ii) for all eigenvalues of absolute value equal to one and multiplicity m, there are m linearly independent eigenvectors $\Bigg\} \Leftrightarrow A \in \mathcal{S}$</td>
<td>Theorem 3-25</td>
</tr>
<tr>
<td>$\forall_{\lambda_i}\ |\lambda_i| < 1 \Leftrightarrow A \in \mathcal{AS}$</td>
<td>Theorem 3-44</td>
</tr>
<tr>
<td rowspan="2">Time-variant
systems

$x(k+1) = A(k)x(k)$</td>
<td>$A(k)$ is regular and reducible to one for all $k \in K$ $\Big\} \Leftrightarrow A(k) \in \mathcal{US}$</td>
<td>Theorem 3-32</td>
</tr>
<tr>
<td>$\forall_{k_0 \in (\alpha,\omega)}\ \exists_{k \in [k_0,\omega)}\ \lambda_{Hmax}(k)=0 \Leftrightarrow A(k) \in \mathcal{AS}$
$(\omega < \infty)$</td>
<td>Lemma 3-41</td>
</tr>
</table>

Table 11

<table>
<tr><td colspan="4" align="center">NECESSARY AND SUFFICIENT
STABILITY CONDITIONS II</td></tr>
<tr>
<td rowspan="3">$x(k+1)=A(k)x(k)+f(k)$</td>
<td>solution x is

stable $\Leftrightarrow A(k) \in \mathcal{S}$

uniformly stable $\Leftrightarrow A(k) \in \mathcal{US}$

asymptotically stable $\Leftrightarrow A(k) \in \mathcal{AS}$

exponentially stable $\Leftrightarrow A(k) \in \mathcal{ES}$</td>
<td></td>
<td>Theorem 4-2</td>
</tr>
<tr>
<td>$\underset{f(.)\in\ell_K^q}{\forall}\ \underset{c>0}{\exists}\ \underset{k\in K}{\forall}\ \|x(k)\| < c \Leftrightarrow A(k) \in \ell^p\mathcal{S}$
$p=q/(q-1)$</td>
<td></td>
<td>Theorem 4-8</td>
</tr>
<tr>
<td>$\underset{q\geq 1}{\forall}\ \underset{f(.)\in\ell_K^q}{\forall}\ \underset{c>0}{\exists}\ \underset{k\in K}{\forall}\ \|x(k)\| < c \Leftrightarrow A(k) \in \mathcal{ES}$</td>
<td></td>
<td>Theorem 4-9</td>
</tr>
<tr>
<td rowspan="2">$x(k+1)=A(k)x(k)+B(k)u(k)$
$y(k)=C(k)x(k)$</td>
<td>$\underset{c>0}{\exists}\ \underset{k\in K}{\forall}\ \sum_{i=\alpha+1}^{k-1} \|\Phi(k,i+1)B(i)\| < c \Leftrightarrow \{A(k),B(k)\}$ is BIBS-st.</td>
<td></td>
<td>Theorem 4-21</td>
</tr>
<tr>
<td>$\underset{c>0}{\exists}\ \underset{k\in K}{\forall}\ \sum_{i=\alpha+1}^{\omega-1} \|G(k,i)\| < c \Leftrightarrow \{A(k),B(k),C(k)\}$ is BIBO-stable</td>
<td></td>
<td>Theorem 4-20</td>
</tr>
</table>

<u>Table 12</u>

Relations between Internal and External Stability

for systems {A(k),B(k),C(k)} with bounded matrices

$A(k) \in ES \Rightarrow \{A(k),B(k)\}$ is ℓ^p-BIBS-stable	*Lemma 4-16*
$A(k) \in \ell^p S \Rightarrow \{A(k),B(k)\}$ is ℓ^q-BIBS-stable for q=p/(p-1)	*Theorem 4-15*
$\{A(k),B(k)\}$ is uniformly stabilizable and ℓ^p-BIBS-stable $\Big\} \Rightarrow A(k) \in ES$	*Theorem 5-26*
$\{A(k),B(k)\}$ is ℓ^p-BIBS-stable $\Rightarrow \{A(k),B(k),C(k)\}$ is ℓ^p-BIBO-stable	*Theorem 5-20*
$\{A(k),B(k),C(k)\}$ is uniformly detectable and ℓ^p-BIBO-stable $\Big\} \Rightarrow \Big\{$ $\{A(k),B(k),C(k)\}$ is ℓ^p-BIBS-stable	*Theorem 5-21*
$A(k) \in ES \Rightarrow \{A(k),B(k),C(k)\}$ is ℓ^p-BIBO-stable	*Theorem 4-22*
$\{A(k),B(k),C(k)\}$ is uniformly stabilizable and uniformly detectable and ℓ^p-BIBO-stable $\Big\} \Rightarrow A(k) \in ES$	*Theorem 5-24*

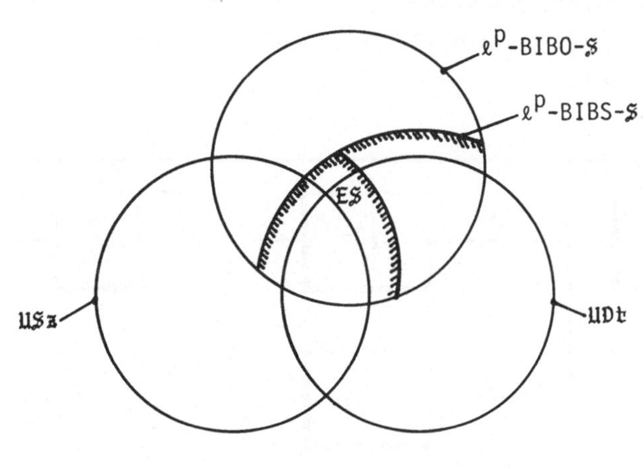

2 Mathematical Description of Discrete-Time Systems

2.1 STATE EQUATION

The internal time response of a discrete-time linear
dynamical system of order n can be described by a system of
n difference equations of order one, the *state equation*,

$$x(k+1) = A(k)x(k) + f(k), \qquad (2.1)$$

where $x \in \mathbb{C}^n$ is the *state vector*, $f \in \mathbb{C}^n$ the *forcing vector*,
$A \in \mathbb{C}^{n \times n}$ the *system matrix*, and $k \in K \subset \mathbb{Z}$ is the *discrete-
time variable*. If the forcing vector f is generated only by
one *input variable* u, then the forcing vector in (2.1) is
$f(k) = b(k)u(k)$, where $b(k) \in \mathbb{C}^n$. In case of m input variables
$u \in R^m$, the forcing vector is $f(k) = B(k)u(k)$, where $B \in \mathbb{C}^{n \times m}$.

The internal response of the system is coupled with the
external world by the *output equation*

$$y(k) = C(k)x(k), \qquad (2.2)$$

where $y \in R^r$ is the *output vector*, and $C \in \mathbb{C}^{r \times n}$ the *output
matrix*. The system structure for the forcing vector $B(k)u(k)$
is sketched in Fig.2/1, where z^{-1} is the delay operator:
$z^{-1}x(k+1) = x(k)$.

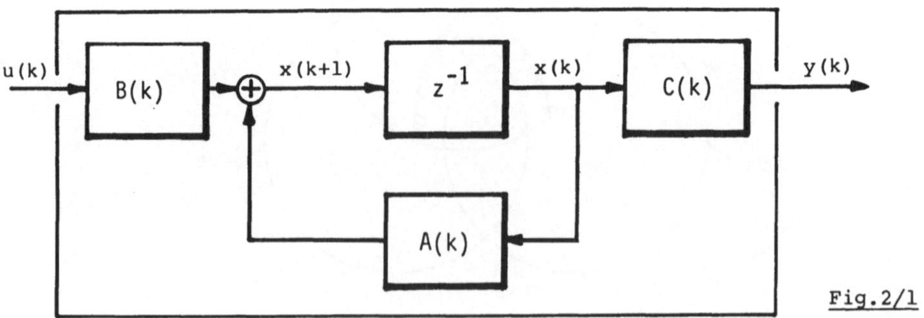

Fig.2/1

The input vector u and the output vector y are elements of real vector spaces, because they are the connections of the system to the real world. If the elements of the state vector x are e.g. physical quantities, the state vector x is an element of a real vector space too. But if the mathematical description is transformed by a state transformation $x = T\hat{x}$, such that the system matrix A becomes e.g. diagonal form, then in case of complex eigenvalues T is an element of $\mathbb{C}^{n \times n}$ and therefore $\hat{x} = T^{-1}x \in \mathbb{C}^n$, $T^{-1}AT \in \mathbb{C}^{n \times n}$, $T^{-1}B \in \mathbb{C}^{n \times m}$, and $CT^{-1} \in \mathbb{C}^{r \times n}$.

The state vector x in equation (2.1) is defined in such a way, that knowledge of the *initial state* $x_o := x(k_o)$ in the initial time point k_o is sufficient for a unique determination of the state vector $x(k)$ for all $k \geq k_o$, if the forcing sequence $f(.) := \{f(k_o), f(k_o+1), \ldots\}$ is known. For the time interval

$$K := (\alpha, \omega), \quad -\infty \leq \alpha < \omega \leq +\infty, \qquad (2.3)$$

with the lowest bound α and the highest bound ω, the solution of the state equation (2.1) is given in the following lemma:

2-1 LEMMA: The solution of the state equation (2.1) is

$$x(k) = x(k; k_o, x_o, f(.))$$

$$= \Phi(k, k_o)x_o + \sum_{i=k_o}^{k-1} \Phi(k, i+1)f(i); \quad k, k_o \in K \quad (2.4)$$

where $\Phi(k, k_o)$ is the *state transition matrix*

$$\Phi(k, k_o) := A(k-1)A(k-2)\ldots A(k_o); \quad k \geq k_o, \qquad (2.5a)$$

$$\Phi(i, i) := I. \qquad (2.5b)$$

PROOF: Putting the solution (2.4) into the state equation (2.1) yields

$$x(k+1) = x(k+1;k_o,x_o,f(.))$$

$$= \Phi(k+1,k_o)x_o + \sum_{i=k_o}^{k} \Phi(k+1,i+1)f(i)$$

$$= A(k)\Phi(k,k_o)x_o + A(k)\sum_{i=k_o}^{k-1} \Phi(k,i+1)f(i) + f(k)$$

$$= A(k)x(k) + f(k).$$

On the other side the solution (2.4) is obtained successively from (2.1) by

$$x(k_o+1) = A(k_o)x(k_o) + f(k_o),$$

$$x(k_o+2) = A(k_o+1)x(k_o+1) + f(k_o+1)$$

$$= A(k_o+1)A(k_o)x_o + A(k_o+1)f(k_o) + f(k_o+1)$$

$$= \Phi(k_o+2,k_o)x_o + \sum_{i=k_o}^{k_o+1} \Phi(k_o+2,i+1)f(i),$$

etc. ∎

If the system matrix A is *time-invariant*, the state transition matrix is

$$\Phi(k,k_o) = A^{k-k_o}, \tag{2.6}$$

and the state equation

$$x(k+1) = A\, x(k) + f(k) \tag{2.7}$$

has for the initial time $k_o=0$ the solution

$$x(k) = A^k x_o + \sum_{i=o}^{k-1} A^{k-i-1}f(i). \tag{2.8}$$

2.2 PROPERTIES OF THE TRANSITION MATRIX

For the free system

$$x(k+1) = A(k)x(k) \qquad (2.9)$$

the solution for $x(k_o)=x_o$ is obtained from Lemma 2-1 by

$$x(k) = \Phi(k,k_o)x_o \ ; \ k,k_o \in K. \qquad (2.10)$$

The property

$$\boxed{\Phi(k,k_1)\Phi(k_1,k_o) = \Phi(k,k_o) \ ; \ k,k_1,k_o \in K,} \qquad (2.11)$$

immediately follows from the definition (2.5) of the state transition matrix, because

$$\Phi(k,k_1)\Phi(k_1,k_o) = [A(k-1)\ldots A(k_1)]\cdot[A(k_1-1)\ldots A(k_o)]$$

$$= \Phi(k,k_o).$$

From the definition equations (2.5) the following matrix difference equations are additionally obtained, which the state transition matrix must satisfy for the 'initial value' $\Phi(k_o,k_o)=I$:

$$\boxed{\begin{aligned} \Phi(k+1,k_o) &= A(k)\Phi(k,k_o) \ ; \ k,k_o \in K. &\qquad (2.12)\\[1em] \Phi(k,k_o-1) &= \Phi(k,k_o)A(k_o-1) \ ; \ k,k_o-1 \in K. &\qquad (2.13) \end{aligned}}$$

2.3 Lagrange Identity and Green's Formula for Difference Equations

Multiplying the state equation (2.9) for the free system from the left by the complex conjugate transposed vector $x^*(k+1):=\bar{x}^T(k+1)$, one obtains

$$x^*(k+1)x(k+1) = x^*(k)A^*(k)A(k)x(k) \ , \qquad (2.14)$$

where $A^*(k):=\bar{A}^T(k)$. With the hermitian matrix

$$H(k) := A^*(k)A(k) \qquad (2.15)$$

equation (2.14) can be written

$$\boxed{x^*(k+1)x(k+1) = x^*(k)H(k)x(k) \ ,} \qquad (2.16)$$

where the right hand side is a hermitian quadratic form. If $A(k)$ is regular, the hermitian matrix $H(k)$ is positive definite, otherwise positive semidefinite. Depending on a similar identity for differential equations [4], the equation (2.16) is called *LAGRANGE Identity*.

With the help of the LAGRANGE Identity (2.16) the following equations for k_o, $k_o+1,\ldots,$ k are obtained

$$x^*(k_o+1)x(k_o+1) - x^*(k_o)x(k_o) = x^*(k_o)[H(k_o)-I]x(k_o) \ ,$$

$$x^*(k_o+2)x(k_o+2) - x^*(k_o+1)x(k_o+1) = x^*(k_o+1)[H(k_o+1)-I]x(k_o+1),$$

$$\vdots$$

$$x^*(k)\,x(k) \ - \ x^*(k-1)\,x(k-1) = x^*(k-1)[H(k-1)-I]x(k-1) \ .$$

The so called *GREEN's Formula* is received by addition of the above equations:

$$x^*(k)x(k) - x^*(k_O)x(k_O) = \sum_{i=k_O}^{k-1} x^*(i)[H(i)-I]x(i) . \qquad (2.17)$$

With the solution (2.10) of the free system (2.9) GREEN's Formula leads to

$$x^*(k_O)\Phi^*(k,k_O)\Phi(k,k_O)x(k_O) - x^*(k_O)Ix(k_O) =$$

$$= \sum_{i=k_O}^{k-1} x^*(k_O)\Phi^*(i,k_O)[H(i)-I]\Phi(i,k_O)x(k_O)$$

for every $x(k_O)\in\mathbb{C}^n$, hence it follows

$$\Phi^*(k,k_O)\Phi(k,k_O) - I = \sum_{i=k_O}^{k-1} \Phi^*(i,k_O)[H(i)-I]\Phi(i,k_O) . \qquad (2.18)$$

If $H(i)-I\equiv O$ is valid, that is if $A^*(i)A(i)\equiv I$, thus $A(i)$ is an unitary matrix, equation (2.18) gives

$$\Phi^*(k,k_O)\Phi(k,k_O) = I . \qquad (2.19)$$

2.4 Estimations for the Norm of the State Transition Matrix

Since $H(k)$ is a positive semidefinite or positive definite hermitian matrix, all eigenvalues of $H(k)$ are real and non negative. The least and greatest of the eigenvalues are, respectively [17],

$$\lambda_{Hmin}(k) := \min_{x^*x>0} \frac{x^*H(k)x}{x^*x}$$

$$= \min_{\|x\|=1} x^*H(k)x , \qquad (2.20)$$

$$\lambda_{Hmax}(k) := \max_{x^*x>0} \frac{x^*H(k)x}{x^*x}$$

$$= \max_{\|x\|=1} x^*H(k)x , \qquad (2.21)$$

such that

$$\lambda_{Hmin}(k)x^*x \le x^*H(k)x \le \lambda_{Hmax}(k)x^*x \ , \ k \in K, \ x \in \mathbb{C}^n. \quad (2.22)$$

Putting the LAGRANGE Identity (2.16) into this inequality, one gets

$$\lambda_{Hmin}(k)x^*(k)x(k) \le x^*(k+1)x(k+1) \le \lambda_{Hmax}(k)x^*(k)x(k) \quad (2.23)$$

and in particular for $k=k_o$

$$\lambda_{Hmin}(k)x^*(k_o)x(k_o) \le x^*(k_o+1)x(k_o+1) \le \lambda_{Hmax}(k_o)x^*(k_o)x(k_o),$$
$$(2.24)$$

and for $k=k_o+1$

$$\lambda_{Hmin}(k_o+1)x^*(k_o+1)x(k_o+1) \le x^*(k_o+2)x(k_o+2) \le$$
$$\le \lambda_{Hmax}(k_o+1)x^*(k_o+1)x(k_o+1). \quad (2.25)$$

With (2.24) the inequality (2.25) yields

$$\lambda_{Hmin}(k_o)\lambda_{Hmin}(k_o+1)x^*(k_o)x(k_o) \le x^*(k_o+2)x(k_o+2) \le$$
$$\le \lambda_{Hmax}(k_o)\lambda_{Hmax}(k_o+1)x^*(k_o)x(k_o). \quad (2.26)$$

In general one gets for the solution $x(k)$ of $x(k+1)=A(k)x(k)$ the estimation

$$\boxed{\begin{aligned} \prod_{i=k_o}^{k-1} \lambda_{Hmin}(i)x^*(k_o)x(k_o) &\le x^*(k)x(k) \le \\ &\le \prod_{i=k_o}^{k-1} \lambda_{Hmax}(i)x^*(k_o)x(k_o) \end{aligned}} \quad (2.27)$$

resp. for $x_o=x(k_o)$ and $x(k)=\Phi(k,k_o)x_o$:

$$\prod_{i=k_o}^{k-1} \lambda_{Hmin}(i)x_o^*x_o \le x_o^*\Phi^*(k,k_o)\Phi(k,k_o)x_o \le$$
$$\le \prod_{i=k_o}^{k-1} \lambda_{Hmax}(i)x_o^*x_o. \quad (2.28)$$

With the usual definition of a matrix norm

$$\|M\| := \max_{\|v\|=1} \|Mv\| \tag{2.29}$$

and since all $\lambda_H(i)$ are non negative, inequality (2.28) yields

$$\prod_{i=k_o}^{k-1} \lambda_{Hmin}(i) \leq \|\Phi(k,k_o)\|^2 \leq \prod_{i=k_o}^{k-1} \lambda_{Hmax}(i) \; , \tag{2.30}$$

hence this bounds for the norm of the state transition matrix:

$$\prod_{i=k_o}^{k-1} \lambda_{Hmin}^{1/2}(i) \leq \|\Phi(k,k_o)\| \leq \prod_{i=k_o}^{k-1} \lambda_{Hmax}^{1/2}(i) . \tag{2.31}$$

From (2.31) follows immediately

$$\exp\left[\sum_{i=k_o}^{k-1} \ln \lambda_{Hmin}^{1/2}(i)\right] \leq \|\Phi(k,k_o)\| \leq \exp\left[\sum_{i=k_o}^{k-1} \ln \lambda_{Hmax}^{1/2}(i)\right] . \tag{2.32}$$

If the system matrix A(k) is regular for all $k \in K$ (for instance this is the case in sampled-data systems), then the hermitian matrix H(k) is positive definite, thus $\lambda_{Hmin}(k) > 0$, and inequality (2.31) yields this estimation for the norm of the state transition matrix:

$$\prod_{i=k_o}^{k-1} \lambda_{Hmax}^{-1/2}(i) \leq \|\Phi(k,k_o)\|^{-1}$$

$$= \|\Phi^{-1}(k,k_o)\| \leq \prod_{i=k_o}^{k-1} \lambda_{Hmin}^{-1/2}(i) \; . \tag{2.33}$$

The next very simple estimation

$$\|\Phi(k,k_o)\| = \|A(k-1)A(k-2)\ldots A(k_o)\|$$

$$\leq \|A(k-1)\|\,\|A(k-2)\|\ldots\|A(k_o)\|$$

$$= \prod_{i=k_o}^{k-1} \|A(i)\| \; , \tag{2.34a}$$

follows from the definition equation (2.5) of the state transition matrix $\Phi(k,k_o)$, thus

$$\boxed{\|\Phi(k,k_o)\| \leq \exp\left[\sum_{i=k_o}^{k-1} \ln\|A(i)\|\right] .} \tag{2.34}$$

If the system matrix A is *time-invariant*, then from (2.31), (2.32), (2.34), resp., the following boundaries for the transition matrix $\Phi(k,0)=A^k$ are valid

$$\lambda_{Hmin}^{k/2} \leq \|A^k\| \leq \lambda_{Hmax}^{k/2} \; , \tag{2.35}$$

$$\exp\left[\frac{k}{2} \ln \lambda_{Hmin}\right] \leq \|A^k\| \leq \exp\left[\frac{k}{2} \ln \lambda_{Hmax}\right] , \tag{2.36}$$

$$\|A^k\| \leq \|A\|^k = \exp\left[k\cdot\ln \|A\|\right] . \tag{2.37}$$

Let $\sigma(A)$ be the spectrum of A, that is the set of all eigenvalues of A, then the following lemma for time-invariant systems is valid.

2-2   **LEMMA:** Let A be *time-invariant*. In order that the inequality

$$\|A^k\| \leq c\,a^k = c\,e^{k\,\ln a}, \quad k \geq 0 , \tag{2.38}$$

holds for some $c\geq 1$ and $a>0$, it is necessary that

$$\lambda \in \sigma(A) \;\Rightarrow\; |\lambda| \leq a , \tag{2.39}$$

and it is sufficient that

$$\lambda \in \sigma(A) \;\Rightarrow\; |\lambda| < a . \tag{2.40}$$

$\textsc{Proof}$: First it is shown that if (2.30) holds and $|\lambda|>a$, then $(A-\lambda I)^{-1}$ exists, i.e., λ is no eigenvalue of A. From (2.38) follows

$$|\lambda|^{-k}\|A^k\| \leq c \ |\lambda|^{-k} \ a^k = c \left(\frac{a}{|\lambda|}\right)^k$$

thus, if $|\lambda|>a$, the matrix sum

$$R(\lambda) := \sum_{i=0}^{\infty} \lambda^{-(i+1)}A^i$$

is finite. Further

$$A\cdot R(\lambda) = R(\lambda)\cdot A$$

$$= \lim_{j\to\infty} \left[\sum_{i=0}^{j} \lambda^{-(i+1)}A^{i+1} \right]$$

$$= \lim_{j\to\infty} \left[\lambda \sum_{i=0}^{j} \lambda^{-(i+2)}A^{i+1} \right]$$

$$= \lim_{j\to\infty} \left[\lambda \left(\sum_{i=0}^{j} \lambda^{-(i+1)}A^i \right) - A^0 + \lambda^{-(j+1)}A^{j+1} \right]$$

$$= \lambda \ R(\lambda) - I \ ,$$

i.e.

$$(\lambda I - A) \ R(\lambda) = R(\lambda)(\lambda I - A) = I \ ,$$

thus the inverse $(\lambda I-A)^{-1}$ exists and λ is no eigenvalue of A. Now the second part of the lemma is proved. If λ_{max} is the eigenvalue of A with maximum value, then the geometric series

$$(zI - A)^{-1} = \frac{1}{z}(I - \frac{1}{z}A)^{-1}$$

$$= \frac{1}{z}(I + Az^{-1} + A^2z^{-2} +...)$$

$$= \sum_{i=0}^{\infty} A^i z^{-(i+1)}$$

converge for $|z|>|\lambda_{max}|$. By assistance of the CAUCHY integral formula

$$A^k = \frac{1}{2\pi i} \int_R (zI-A)^{-1} z^k dz \ ,$$

A^k can be computed. From this formula the following estimation is obtained

$$\|A^k\| \leq \frac{1}{2\pi} \int_R \|(zI-A)^{-1}\| \, |z^k| \, |dz|$$

$$\leq \left[\frac{1}{2\pi} \int_R \|(zI-A)^{-1}\| \, |dz| \right] a^k ,$$

i.e. with

$$c := \frac{1}{2\pi} \int_R \|(zI-A)^{-1}\| \, |dz|$$

the inequality (2.38) is received.∎

Additionally bounds for the norm of the transition matrix can be obtained in the following way. Let $w(.)$ be a positive sequence ($w(i)>0$), then

$$\|\Phi(k,k_1)\| \sum_{i=k_O}^{k-1} w(i) = \sum_{i=k_O}^{k-1} \|\Phi(k,k_1)\| w(i)$$

$$\leq \sum_{i=k_O}^{k-1} \|\Phi(k,i)\| \, \|\Phi(i,k_1)\| \, w(i) \ ,$$

thus

$$\|\Phi(k,k_1)\| \leq \left[\sum_{i=k_O}^{k-1} w(i) \right]^{-1} \sum_{i=k_O}^{k-1} \|\Phi(k,i)\| \, \|\Phi(i,k_1)\| w(i) \qquad (2.41)$$

for all $k>k_O, k_1 \in K$. In the case of $k_O<k_1$ the state transition matrix $\Phi(i,k_1)$ for $i<k_1$ is defined as

$$\Phi(i,k_1) := A^{-1}(i) A^{-1}(i+1) \ldots A^{-1}(k_1-1) \ ,$$

i.e. the needed system matrices must be regular!

Specially for $w(i)\equiv 1$ the following estimation is obtained from (2.41):

$$\|\Phi(k,k_1)\| \leq \frac{1}{k-k_o} \sum_{i=k_o}^{k-1} \|\Phi(k,i)\|\,\|\Phi(i,k_1)\| \quad ; \quad k>k_o, k_1\in K. \qquad (2.42)$$

Further for $w(i) = \|\Phi(i,k_1)\|^{-1}$ one gets from (2.41) the estimation

$$\|\Phi(k,k_1)\| \leq \left[\sum_{i=k_o}^{k-1} \|\Phi(i,k_1)\|^{-1}\right]^{-1} \sum_{i=k_o}^{k-1} \|\Phi(k,i)\| \quad ; \quad k>k_o, k_1\in K, \qquad (2.43)$$

and in particular for $j\geq k_1$:

$$\left[\sum_{i=k_o}^{j-1} \|\Phi(j,i)\|\right]^{-1} \leq \|\Phi(j,k_1)\|^{-1}\left[\sum_{i=k_o}^{j-1} \|\Phi(i,k_1)\|^{-1}\right]^{-1}. \qquad (2.44)$$

For $j=k_1, k_1+1, \ldots, k-1$ the inequality (2.44) yields

$$\left[\sum_{i=k_o}^{k_1-1} \|\Phi(k_1,i)\|\right]^{-1} \leq \|\Phi(k_1,k_1)\|^{-1}\left[\sum_{i=k_o}^{k_1-1} \|\Phi(i,k_1)\|^{-1}\right]^{-1}, \qquad (2.45)$$

$$\left[\sum_{i=k_o}^{(k_1+1)-1} \|\Phi(k_1+1,i)\|\right]^{-1} \leq \|\Phi(k_1+1,k_1)\|^{-1}\left[\sum_{i=k_o}^{k_1} \|\Phi(i,k_1)\|^{-1}\right]^{-1}$$

$$\leq \|\Phi(k_1+1,k_1)\|^{-1}\left[\sum_{i=k_o}^{k_1-1} \|\Phi(i,k_1)\|^{-1}\right]^{-1}, \qquad (2.46)$$

$$\vdots$$

$$\left[\sum_{i=k_o}^{(k-1)-1} \|\Phi(k,i)\|\right]^{-1} \leq \|\Phi(k-1,k_1)\|^{-1}\left[\sum_{i=k_o}^{k_1-1} \|\Phi(i,k_1)\|^{-1}\right]^{-1}. \qquad (2.47)$$

Summation of (2.45) to (2.47) gives as a result

$$\sum_{j=k_1}^{k-1} \left[\sum_{i=k_o}^{j-1} \|\Phi(j,i)\|\right]^{-1} \leq \left[\sum_{j=k_1}^{k-1} \|\Phi(j,k_1)\|^{-1}\right]\left[\sum_{i=k_o}^{k_1-1} \|\Phi(i,k_1)\|^{-1}\right]^{-1},$$

i.e.

$$\left[\sum_{i=k_1}^{k-1} \|\Phi(i,k_1)\|^{-1}\right]^{-1} \leq \left[\sum_{i=k_O}^{k_1-1} \|\Phi(i,k_1)\|^{-1}\right]^{-1} \left\{\sum_{j=k_1}^{k-1}\left[\sum_{i=k_O}^{j-1} \|\Phi(j,i)\|\right]^{-1}\right\}^{-1}.$$

$$\tag{2.48}$$

Putting (2.48) into (2.43) finally yields the estimation

$$\|\Phi(k,k_1)\| \leq \left[\sum_{i=k_O}^{k-1} \|\Phi(k,i)\|\right]\left[\sum_{i=k_O}^{k_1-1} \|\Phi(i,k_1)\|^{-1}\right]^{-1} \times$$

$$\times \left\{\sum_{j=k_1}^{k-1}\left[\sum_{i=k_O}^{j-1} \|\Phi(j,i)\|\right]^{-1}\right\}^{-1}.$$

$$\tag{2.49}$$

Now for $w(i)=\|\Phi(i,k_1)\|^{-p}$ and $p>1$ inequality (2.41) produces

$$\|\Phi(k,k_1)\| \leq \left[\sum_{i=k_O}^{k-1} \|\Phi(i,k_1)\|^{-p}\right]^{-1} \sum_{i=k_O}^{k-1} \|\Phi(k,i)\|\,\|\Phi(i,k_1)\|^{1-p}$$

$$\tag{2.50}$$

for $k>k_O,k_1 \in K$. By assistance of the HÖLDER inequality the right hand side sum in (2.50) can be converted in this way $(1/q + 1/p = 1)$:

$$\sum_{i=k_O}^{k-1} \|\Phi(k,i)\|\,\|\Phi(i,k_1)\|^{1-p} \leq \left[\sum_{i=k_O}^{k-1} \|\Phi(k,i)\|^{p}\right]^{\frac{1}{p}}\left[\sum_{i=k_O}^{k-1} \|\Phi(i,k_1)\|^{q(1-q)}\right]^{\frac{1}{q}}$$

$$= \left[\sum_{i=k_O}^{k-1} \|\Phi(k,i)\|^{p}\right]^{\frac{1}{p}}\left[\sum_{i=k_O}^{k-1} \|\Phi(i,k_1)\|^{-p}\right]^{1-\frac{1}{p}}.$$

$$\tag{2.51}$$

Insert (2.51) into (2.50) produces

$$\|\Phi(k,k_1)\| \leq \left[\sum_{i=k_O}^{k-1} \|\Phi(k,i)\|^{p}\right]^{\frac{1}{p}}\left[\sum_{i=k_O}^{k-1} \|\Phi(i,k_1)\|^{-p}\right]^{-\frac{1}{p}};$$

$$\tag{2.52}$$

$$k > k_O,k_1 \in K; \quad p \geq 1$$

which generalizes (2.43).

The estimation

$$\|\Phi(k,k_1)\| \leq \left[\sum_{i=k_O}^{k-1} \|\Phi(k,i)\|^p\right]^{\frac{1}{p}} \left[\sum_{i=k_O}^{k-1} \|\Phi(i,k_1)\|^{-p}\right]^{-\frac{1}{p}} \times$$

$$\times \left\{\sum_{j=k_1}^{k-1} \left[\sum_{i=k_O}^{j-1} \|\Phi(j,i)\|^p\right]^{-1}\right\}^{-\frac{1}{p}} \qquad (2.53)$$

follows from (2.52) in the same way as the estimation (2.49) follows from (2.43).

For the determinant of the transition matrix $\Phi(k,k_O)$ from the definition formula (2.5) follows

$$\det \Phi(k,k_O) = \det [A(k-1)A(-2)\ldots A(k_O)]$$

$$= \prod_{i=k_O}^{k-1} \det A(i) \qquad (2.54)$$

$$= \exp\left[\sum_{i=k_O}^{k-1} \ln[\det A(i)]\right] , \qquad (2.55)$$

where (2.55) is valid only if $\det A(i) > 0$.

If the discrete-time system results from sampling a continuous-time system, then

$$A(i) = \Phi_F(i\tau+\tau, i\tau) ,$$

where Φ_F is the state transition matrix of the continuous-time system

$$\dot{x}(t) = F(t) x(t)$$

and τ is the constant sampling period. For the determinant of the state transition matrix of the continuous-time system the JACOBI - LIOUVILLE - Identity [7] is valid:

$$\det \Phi_F(t,t_o) = \exp\left[\int_{t_o}^{t} \text{trace } F(\tau)\, d\tau\right] ,$$

i.e. especially for $t=(i+1)\tau$ and $t_o=i\tau$

$$\det A(i) = \exp\left[\int_{i\tau}^{(i+1)\tau} \text{trace } F(t)\, dt\right].\tag{2.56}$$

Putting (2.56) into (2.54) yields the result

$$\boxed{\det \Phi(k,k_o) = \exp\left[\sum_{i=k_o}^{k-1}\int_{i\tau}^{(i+1)\tau} \text{trace } F(t)\, dt\right].}\tag{2.57}$$

An important identity shall now be derived. Let Φ_A, resp. Φ_B the state transition matrices of the discrete-time systems $x(k+1)=A(k)x(k)$, $x(k+1)=B(k)x(k)$, resp. Then one gets with (2.12) and (2.13):

$$\Phi_A(k,s+1)\ \Phi_B(s+1,k_o) - \Phi_A(k,s)\ \Phi_B(s,k_o) =$$

$$= \Phi_A(k,s+1)\ [B(s)\Phi_B(s,k_o)]-[\Phi_A(k,s+1)A(s)]\Phi_B(s,k_o)$$

$$= \Phi_A(k,s+1)\ [B(s)-A(s)]\Phi_B(s,k_o).\tag{2.58}$$

For $s=k_o$ till $s=k-1$ equation (2.58) produces the following equations

$$\Phi_A(k,k_o+1)\Phi_B(k_o+1,k_o)-\Phi_A(k,k_o)=\Phi_A(k,k_o+1)[B(k_o)-A(k_o)],\tag{2.59}$$

$$\Phi_A(k,k_o+2)\Phi_B(k_o+2,k_o) - \Phi_A(k,k_o+1)\Phi_B(k_o+1,k_o) =$$

$$= \Phi_A(k,k_o+2)[B(k_o+1)-A(k_o+1)]\Phi_B(k_o+1,k_o) ,$$

$$\vdots$$

$$\Phi_B(k,k_o)-\Phi_A(k,k-1)\Phi_B(k-1,k_o)=[B(k-1)-A(k-1)]\Phi_B(k-1,k_o).\tag{2.60}$$

Adding up (2.59) to (2.60) yields the identity

$$\Phi_B(k,k_o) - \Phi_A(k,k_o) = \sum_{i=k_o}^{k-1} \Phi_A(k,i+1)[B(i)-A(i)]\Phi_B(i,k_o) \ . \qquad (2.61)$$

Let Δ be a subinterval of K and

$$M := \max \ [\|\Phi_A(k,k_o)\| : k,k_o \in \Delta] \ ,$$

then the identity (2.61) yields

$$\|\Phi_B(k,k_o)\| \le M + \sum_{i=k_o}^{k-1} M\|B(i)-A(i)\| \ \|\Phi_B(i,k_o)\| \ . \qquad (2.62)$$

For the further evaluation of the estimation (2.62) the
following lemma is needed:

2-3   LEMMA: If (i) $a(i) \ge 0$ and $b(i) \ge 0$,

(ii) $c > 0$,

(iii) $a(k_o) \le c$,   (2.63)

(iv) $a(k) \le c + \sum_{i=k_o}^{k-1} a(i)b(i)$ for   (2.64)

$k=k_o+1, \ k_o+2, \ldots,$

then the following inequality is valid for
$k=k_o+1, \ k_o+2, \ldots:$

$$a(k) \le c \prod_{i=k_o}^{k-1} [1+b(i)]$$

$$\le c \ \exp\left[\sum_{i=k_o}^{k-1} b(i)\right] . \qquad (2.65)$$

Lemma 2-3 is the discrete analogue to the well known GRONWALL-
BELLMAN-Lemma for continuous functions.

PROOF: From the assumptions (2.63) and (2.64) one gets successively

$$a(k_o) \quad \le c \ ,$$

$$a(k_o+1) \leq c + a(k_o)b(k_o)$$

$$\leq c[1 + b(k_o)],$$

$$a(k_o+2) \leq c + a(k_o)b(k_o) + a(k_o+1)b(k_o+1)$$

$$\leq c[1 + b(k_o)] + c[1 + b(k_o)]b(k_o+1)$$

$$= c[1 + b(k_o)][1 + b(k_o+1)] ,$$

$$\vdots$$

$$a(k) \leq c \prod_{i=k_o}^{k-1} [1 + b(i)] . \tag{2.66}$$

On the other side is for $b(i) \geq 0$

$$1 + b(i) \leq 1 + b(i) + 1/2 \, b^2(i) + \ldots = e^{b(i)},$$

hence

$$\ln [1 + b(i)] \leq b(i). \tag{2.67}$$

Further

$$\prod_{i=k_o}^{k-1} [1 + b(i)] = \exp \left[\ln \prod_{i=k_o}^{k-1} [1 + b(i)] \right]$$

$$= \exp \left[\sum_{i=k_o}^{k-1} \ln [1 + b(i)] \right] \tag{2.68}$$

is valid. Putting (2.67) into (2.68), and the result into (2.66), yields finally the assertion (2.65). ∎

Now with Lemma 2-3 one obtains from (2.62) this estimation

$$\| \Phi_B(k,k_o) \| \leq M \exp \left[\sum_{i=k_o}^{k-1} M \| B(i) - A(i) \| \right] ; \quad k, k_o \in \Delta \subset K. \tag{2.69}$$

Therewith (2.61) yields for $k, k_o \in \Delta$:

$$\| \Phi_B(k,k_o) - \Phi_A(k,k_o) \| \leq \sum_{i=k_o}^{k-1} M \, \| B(i) - A(i) \| \, \| \Phi_B(i,k_o) \|$$

$$\leq M^2 \sum_{\Delta} \| B(i) - A(i) \| \exp \left[\sum_{\Delta} M \| B(i) - A(i) \| \right] . \quad (2.70)$$

From (2.70) follows

<u>2-4</u>

> LEMMA: For every $\varepsilon > o$ there exists a $\delta > o$ such that the implication
>
> $$\| B(i) - A(i) \| < \delta \quad \Rightarrow \quad \| \Phi_B(k,k_o) - \Phi_A(k,k_o) \| < \varepsilon$$
>
> is valid.

3 Stability of Free Discrete-Time Systems

This chapter is dealed with the asymptotic behaviour of the free linear time-varying discrete-time system

$$x(k+1) = A(k) \ x(k), \qquad (3.1)$$

i.e., with the behaviour of the solutions of (3.1)

$$x(k;k_o,x_o) = \Phi(k,k_o) \ x_o \ , \quad k \in K = (\alpha,\omega), \qquad (3.2)$$

for $k \to \omega$ independent of the initial state x_o, hence with the asymptotic behaviour of the state transition matrix $\Phi(k,k_o)$.

3.1 LJAPUNOW- AND LAGRANGE-STABILITY

In case of stability analysis in the sense of LJAPUNOW it is of interest, whether the new motion $\tilde{x}(k):=x(k;k_o,\tilde{x}_o)$ remains close to the old motion $x(k)=x(k;k_o,x_o)$ for $k,k_o \in K$ after a variation of the initial state $x_o=x(k_o)$ into $\tilde{x}_o$. Here the 'old motion' also can be an equilibrium point x_e. A vector x_e is said to be an *equilibrium point* if once the state vector is equal to x_e it remains equal to x_e for all future time. Hence for an equilibrium point

$$x_e = A(k) \ x_e \qquad (3.3)$$

is valid, thus

$$[A(k) - I] \ x_e = O \ ,$$

i.e.

$$x_e \in \text{kernel} \ [A(k) - I]. \qquad (3.4)$$

EXAMPLE 3.1: For the linear time-variant discrete-time
system

$$x(k+1) = \begin{bmatrix} 1 & a(k) \\ 0 & 1 \end{bmatrix} x(k) \quad , \quad a(k) \neq 0,$$

all states $[x_1,0]^T$, $x_1 \in \mathbf{C}$, are equilibrium states, since

$$\text{kernel } [A(k) - I] = \left\{ x : \begin{bmatrix} 0 & a(k) \\ 0 & 0 \end{bmatrix} x = 0 \right\}.$$

The possible motions and the equilibrium states are sketched
in Fig.3/1 for a>0.∎

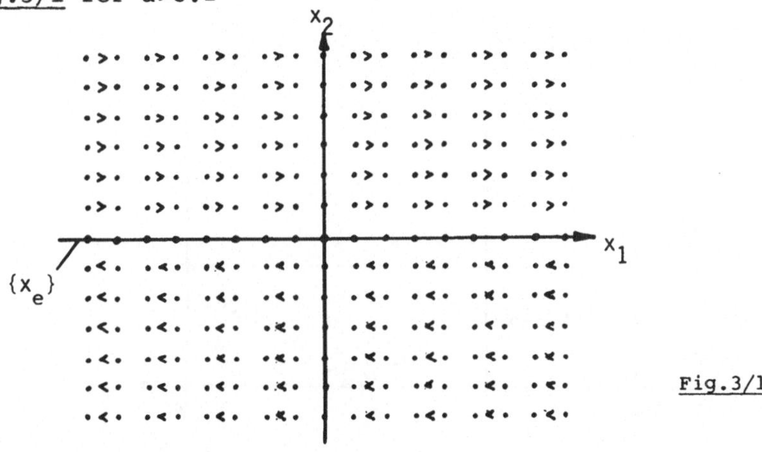

Fig.3/1

The measure for the distance of two points in a vector
space is the euclidian norm $\|x\| := (x^*x)^{1/2}$, if nothing else is
said.

3-1

DEFINITION: A solution x of the system $x(k+1)=A(k)x(k)$
is said to be *stable* (or *LJAPUNOW-stable*),
if for every $\varepsilon>0$ and every $k_o \in K$ exists a
$\delta(k_o,\varepsilon)>0$, such that $\|\tilde{x}_o-x_o\|<\delta$ implies

$$\|\tilde{x}(k;k_o,\tilde{x}_o) - x(k;k_o,x_o)\| < \varepsilon \tag{3.5}$$

for all $k \in [k_o,\omega)$.

The contents of Definition 3-1 are visualized by <u>Fig.3/2</u>
and <u>Fig.3/3</u>.

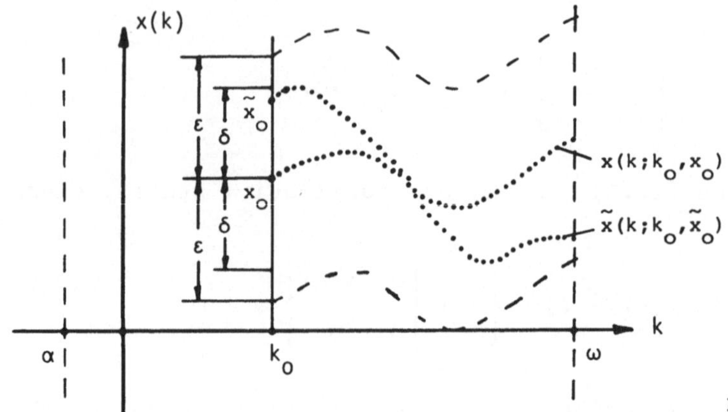

<u>Fig.3/2</u>

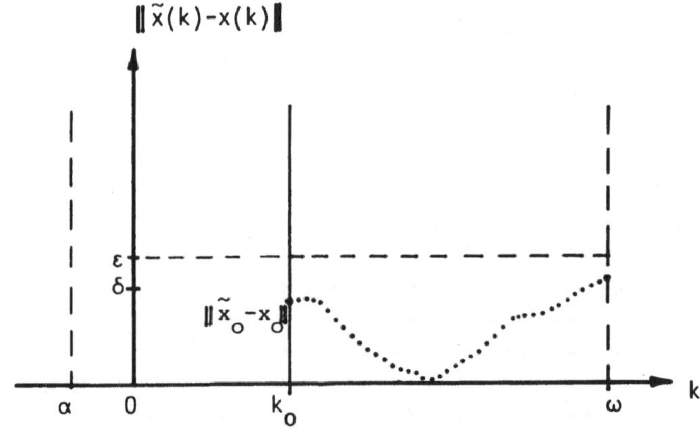

<u>Fig.3/3</u>

From

$$\tilde{x}(k;k_o,\tilde{x}_o) - x(k;k_o,x_o) = \Phi(k,k_o)(\tilde{x}_o - x_o) \tag{3.6}$$

the estimation

$$\left\| \tilde{x}(k;k_o,\tilde{x}_o) - x(k;k_o,x_o) \right\| \leq \left\| \Phi(k,k_o) \right\| \left\| \tilde{x}_o - x_o \right\| , \tag{3.7}$$

is following, i.e., the behaviour of the vector difference
norm in (3.7) is characterized by the behaviour of the norm
of the state transition matrix $\Phi(k,k_o)$. For this reason now

the definition:

3-2   DEFINITION: The matrix $A(k)$ is said to be of *class $\mathcal{S}$*, $A(k) \in \mathcal{S}$, if for every $k_o \in K$ exists a positive constant $c(k_o) < \infty$, so that for every $k \in [k_o, \omega)$

$$\| \Phi(k, k_o) \| < c(k_o) \tag{3.8}$$

is valid, i.e., the state transition matrix is bounded.

From Definition 3-2 and estimation (3.7) immediately follows

3-3   LEMMA: If $A(k) \in \mathcal{S}$, then every solution of $x(k+1)=A(k)x(k)$ is stable.

On the other side the following lemma is valid:

3-4   LEMMA: If a solution of $x(k+1)=A(k)x(k)$ is stable, then $A(k) \in \mathcal{S}$.

PROOF: If x is a stable solution of $x(k+1)=A(k)x(k)$, then exists for $\varepsilon=1$ and $k_o \in K$ a $\delta=\delta(k_o)>0$ so that for $k \geq k_o > \alpha$: $\| \tilde{x}_o - x_o \| < \delta$ implies $\| \Phi(k, k_o)(\tilde{x}_o - x_o) \| < 1$, i.e., with $c=1/\delta$ follows (3.8). $\blacksquare$

Lemma 3-3 together with Lemma 3-4 yields:

3-5   LEMMA: If a solution of $x(k+1)=A(k)x(k)$ is stable, then every solution is stable.

Therefore it makes sense to define:

3-6   DEFINITION: The _system_ $x(k+1)=A(k)x(k)$ is said to be _stable_, if every solution x is stable.

Combining Lemma 3-3 and Lemma 3-4 one gets:

3-7   THEOREM: The system $x(k+1)=A(k)x(k)$ is stable, if and only if $A(k) \in \mathcal{S}$.

From the estimations (2.31) and (2.32) for the state transition matrix $\Phi(k,k_o)$ one obtains directly the necessary stability condition in the following Lemma 3-8 and the sufficient stability condition in Lemma 3-9.

3-8   LEMMA: If $A(k) \in \mathcal{S}$ the following two inequalities are valid $(H:=A^*A)$:

$$\lim_{k\to\omega} \prod_{i=k_o}^{k-1} \lambda_{H min}^{\sqrt{2}}(i) < \infty, \tag{3.9}$$

$$\lim_{k\to\omega} \sum_{i=k_o}^{k-1} \frac{1}{2} \ln \lambda_{H min}(i) < \infty. \tag{3.10}$$

3-9   LEMMA: If for all $k \geq k_o$, $k \in K$

$$\prod_{i=k_o}^{k-1} \lambda_{H max}^{\sqrt{2}}(i) < \infty \ , \quad \text{resp.} \tag{3.11}$$

$$\sum_{i=k_o}^{k-1} \frac{1}{2} \ln \lambda_{H max}(i) < \infty \tag{3.12}$$

is valid, then $A(k) \in \mathcal{S}$.

If only the distance of the solution x from the trivial equilibrium state $x_e \equiv 0$ is of interest, then one can use the following definitions.

3-10   **DEFINITION:** The solution x of $x(k+1)=A(k)x(k)$ is said to be *bounded*, or *LAGRANGE-stable*, if for every $k_o \in K$ there exists a positive constant $\mu(k_o) < \infty$ such that for all $k \in [k_o, \omega)$

$$\|x(k; k_o, x_o)\| \leq \mu(k_o) \tag{3.13}$$

is valid.

3-11   **DEFINITION:** The *system* $x(k+1)=A(k)x(k)$ is said to be *LAGRANGE-stable*, if every solution of the system is bounded.

A connection between stability and LAGRANGE-stability is established in the following theorem.

3-12   **THEOREM:** The system $x(k+1)=A(k)x(k)$ is LAGRANGE-stable if and only if $A(k) \in \mathcal{S}$, i.e., if the system is stable.

PROOF: ($\Leftarrow$) If $A(k) \in \mathcal{S}$ the system is naturally LAGRANGE-stable according to Definition 3-2 and Definition 3-10.

($\Rightarrow$) Let $e_1, e_2, \ldots, e_n$ be a basis of the state space, such that every initial state x_o can be represented by

$$x_o = \sum_{i=1}^{n} (x_o^* e_i) e_i .$$

Therefore for every initial state the estimation

$$\left| x_o^* \, e_i \right| \leq \left\| x_o \right\|$$

is valid, and for every solution one has

$$x(k;k_o,x_o) = \Phi(k,k_o) \, x_o$$

$$= \sum_{i=1}^{n} (x_o^* \, e_i) \, \Phi(k,k_o) \, e_i$$

$$= \sum_{i=1}^{n} (x_o^* \, e_i) \, x(k;k_o,e_i).$$

If every solution is bounded, then

$$\left\| x(k;k_o,x_o) \right\| \leq \sum_{i=1}^{n} \left\| x_o \right\| \mu(k_o) \,,$$

i.e., that the zero solution is stable, thus in consequence of Lemma 3-4, $A(k) \in \mathcal{S}$.∎

Hence in case of linear systems $x(k+1)=A(k)x(k)$ there must not be distinguished between LJAPUNOW-stability and LAGRANGE-stability!

3.2 Short Time Boundedness

In this section the problem is considered whether a system response belongs to specified bounds over specified intervals of time.

<u>3-13</u> | **DEFINITION:** The *solution* x of the system $x(k+1)=A(k)x(k)$ is said to be *short time bounded*, if for *given* finite $\delta>0$, $\varepsilon>0$ $(\delta\leq\varepsilon)$, and $N>0$ for every $k_o \in (\alpha,\omega-N)$, $\left\| x_o \right\| < \delta$ implies

$$\|x(k;k_o,x_o)\| < \varepsilon \qquad (3.14)$$

for all $k\in[k_o,k_o+N]$.
The *system* $x(k+1)=A(k)x(k)$ is said to be *short time bounded*, if every solution x is short time bounded.

3-14   **DEFINITION:** The matrix $A(k)$ is said to be of *class $\mathscr{SB}$*, $A(k) \in \mathscr{SB}$, if for *given* $\varepsilon>0$, $\delta>0$, $N>0$, and $k_o\in(\alpha,\omega-N)$

$$\|\Phi(k,k_o)\| \leq \varepsilon/\delta \qquad (3.15)$$

is valid for all $k\in[k_o,k_o+N]$.

Since

$$\|x(k;k_o,x_o)\| \leq \|\Phi(k,k_o)\|\,\|x_o\| \leq \|\Phi(k,k_o)\|\,\delta$$

it immediately follows, that condition (3.14) is fulfilled if and only if (3.15) is valid. So one has

---

3-15   **THEOREM:** The system $x(k+1)=A(k)x(k)$ is short time bounded if and only if $A(k) \in \mathscr{SB}$.

With the estimations (2.31) and (2.32) one obtains the sufficient condition of Theorem 3-16 and the necessary condition of Theorem 3-17.

3-16   **THEOREM:** If for all $k_o\in(\alpha,\omega-N)$

$$\prod_{i=k_o}^{k-1} \lambda_{Hmax}^{1/2}(i) \leq \varepsilon/\delta \qquad (3.16)$$

or

$$\sum_{i=k_o}^{k-1} \ln \lambda_{Hmax}(i) \leq 2 \ln(\varepsilon/\delta) \tag{3.17}$$

is valid for all $k\in[k_o,k_o+N]$, where λ_{Hmax} is the maximum eigenvalue of $H(k)=A^*(k)A(k)$, then $A(k) \in \mathscr{SB}$.

3-17   THEOREM: If $A(k) \in \mathscr{SB}$, then for all $k_o\in(\alpha,\omega-N)$ and $k\in[k_o,k_o+N]$

$$\prod_{i=k_o}^{k-1} \lambda_{Hmin}^{1/2}(i) \leq \varepsilon/\delta \tag{3.18}$$

or

$$\sum_{i=k_o}^{k-1} \ln \lambda_{Hmin}(i) \leq 2 \ln(\varepsilon/\delta) \tag{3.19}$$

is valid.

There are systems for which $A(k)\in\mathscr{SB}$ but $A(k)\notin\mathscr{S}$ and vice versa such that

$$\mathscr{S} \neq \mathscr{SB} \tag{3.20}$$

and the relation in <u>Fig.3/4</u> is valid.

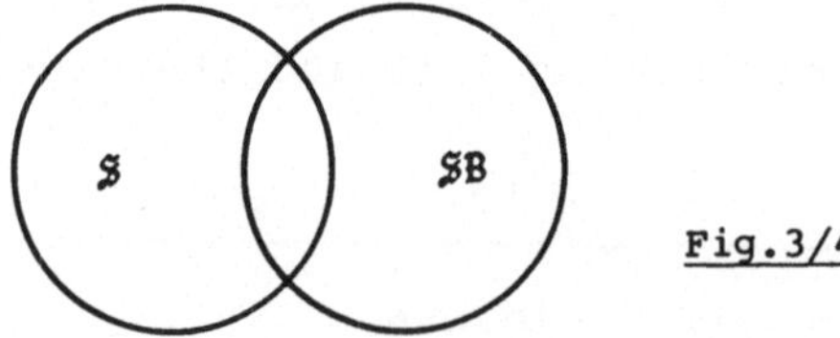

<u>Fig.3/4</u>

EXAMPLE 3.2: For $\delta=0.5$, $\varepsilon=10$, and $N=10$, the linear time-invariant system $x(k+1)=13x(k)$ is short time bounded but not stable on $K=(0,\infty)$ because (see <u>Fig.3/5</u>)

$$\|\Phi(k_o+N,k_o)\| = \|1.3^{10}\| = 13.78... \leq \varepsilon/\delta = 20$$

but

$$\lim_{k \to \infty} \|1.3^{k_o+k}\| \to \infty. \quad \blacksquare$$

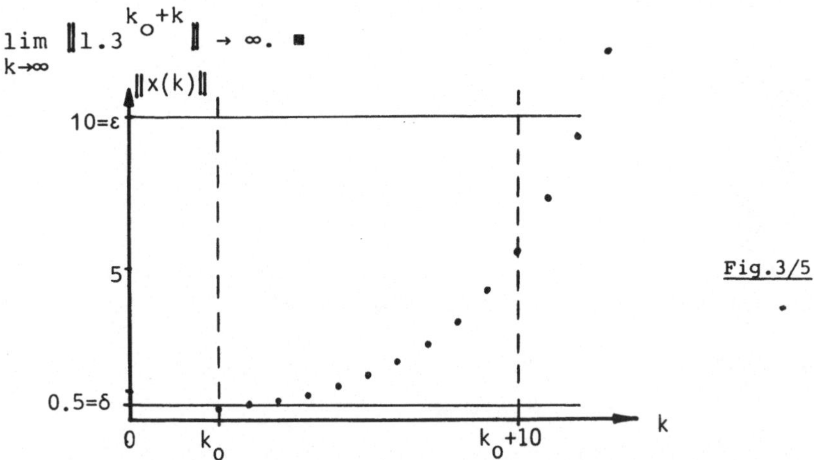

Fig.3/5

EXAMPLE 3.3: For the same δ, ε, and N as in Example 3.2 above, the linear time-variant system

$$x(k+1) = \frac{2.8}{k - 5.5} \, x(k)$$

is not short time bounded but stable on $K=(0,\omega)$, see Fig.3/6. $\blacksquare$

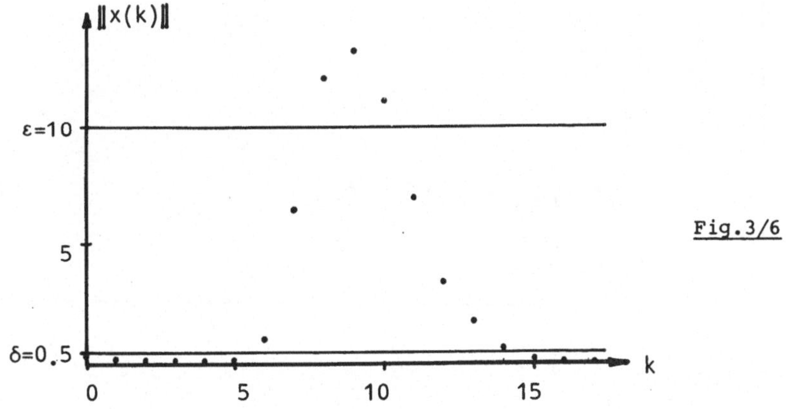

Fig.3/6

If the system is *time-invariant*, i.e., the system matrix A is constant, then one has for $k_o=0$ in Definition 3-14 the condition

$$\|A^k\| \leq \varepsilon/\delta \tag{3.21}$$

for $A \in \mathcal{B}$. Since for all eigenvalues of a matrix A the following bounds are valid [17]

$$\lambda_{A^*A,min}^{1/2} \leq |\lambda_A| \leq \lambda_{A^*A,max}^{1/2} \tag{3.22}$$

and moreover

$$\lambda_{A^k} = \lambda_A^k \tag{3.23}$$

it is by definition of the euclidian norm

$$\left\| A^k \right\| = \lambda_{A^{*k}A^k,max}^{1/2} = \lambda_{A^*A,max}^{k/2}$$
$$\geq \left| \lambda_A^k \right|_{max} = \left| \lambda_A \right|_{max}^k .$$

Thus from Theorem 3-16 follows

3-18   LEMMA: If the system matrix A is constant and $A \in \mathcal{B}$, then for the given δ, ε, and N

$$\left| \lambda_A \right|_{max} \leq (\varepsilon/\delta)^{1/N} \tag{3.24}$$

or

$$\ln\left| \lambda_A \right|_{max} \leq \frac{1}{N} \ln(\varepsilon/\delta) \tag{3.25}$$

is valid.

EXAMPLE 3.4: Indeed for the system in Example 3.2 is

$$\left| \lambda_A \right|_{max} = 1.3 \leq (\varepsilon/\delta)^{1/N} = 20^{.1} = 1.349\ldots \;\blacksquare$$

3.3 Uniform Stability

In Definition 3-2 the constant $c(k_0)$ of the property $\|\Phi(k,k_0)\| < c(k_0)$ for the class $\mathcal{S}$ depends on the initial time k_0. If the constant c is assumed to be independent of k_0, the following definition is necessary:

3-19   **DEFINITION:** The matrix $A(k)$ is said to be of *class* $\mathfrak{U}\mathcal{S}$, $A(k) \in \mathfrak{U}\mathcal{S}$, if a positive constant $c < \infty$ exists so that for every $k_0 \in K$ and every $k \in [k_0, \omega)$

$$\|\Phi(k,k_0)\| < c \tag{3.26}$$

is valid.

A comparison of Definition 3-2 with Definition 3-19 shows immediately that the class $\mathfrak{U}\mathcal{S}$ is a subclass of $\mathcal{S}$:

$$\mathfrak{U}\mathcal{S} \subset \mathcal{S} \ . \tag{3.27}$$

But it is not $\mathfrak{U}\mathcal{S} = \mathcal{S}$! For this purpose see

EXAMPLE 3.5: The system of first order

$$x(k+1) = f(k+1)\, f^{-1}(k)\, x(k) \ , \tag{3.28}$$

where $f(i) > 0$ for $i \in K$, has the 'transition matrix'

$$\Phi(k,k_0) = \prod_{i=k_0}^{k-1} f(i+1) f^{-1}(i)$$

$$= [f(k) f^{-1}(k-1)][f(k-1) f^{-1}(k-2)] \ldots [f(k_0+1) f^{-1}(k_0)]$$

$$= f(k) f^{-1}(k_0) . \tag{3.29}$$

In particular let be $K=[0,\infty)$ and $f(k)=\exp[-k\ \sin^2(k\pi/2)]$.
Then it is

$$\Phi(k,k_O) = \exp[k_O\sin^2(k_O\pi/2)-k\ \sin^2(k\pi/2)]\ ,$$

i.e., for $k_O=0,\ 2,\ 4,\ldots$ and $k\geq k_O$

$$\left\|\Phi(k,O)\right\| = \left|\exp[-k\ \sin^2(k\pi/2)]\right| \leq 1$$

is valid, thus $A(k)\in\$$. But for $k_O=1,\ 3,\ 5,\ldots$ and $k=k_O+1$
the 'transition matrix' is $\Phi(k_O+1,k_O)=\exp(k_O)$, i.e., $A(k)\notin U\$$,
(<u>Fig.3/7</u>).■

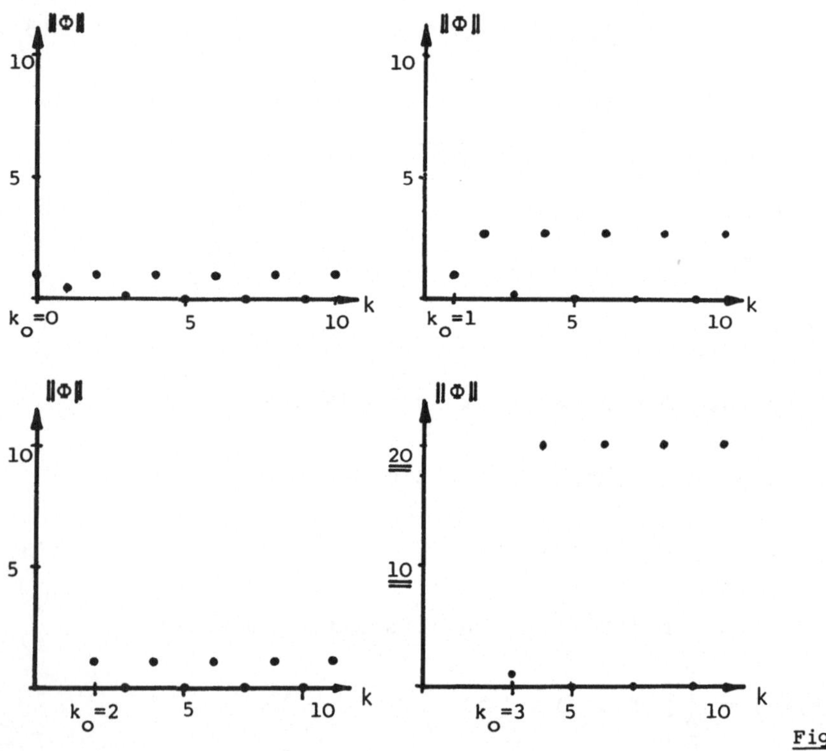

Fig.3/7

EXAMPLE 3.6: Let $K=(0,\infty)$ and

$$f(k) = k^{-2+\cos(k\pi)}$$

for the system (3.28). Since

$$\lim_{k \to \infty} \Phi(k,k_o) = O$$

it is $A(k) \in \mathcal{S}$, whereas $A(k) \notin \mathcal{US}$ because

$$\Phi(k_o+1,k_o) = f(k_o+1)f^{-1}(k_o)$$

$$= (k_o+1)^{-1}\left[(k_o)^{-3}\right]^{-1}$$

$$= k_o^3/(k_o+1)$$

for $k_o=1,\ 3,\ 5,\ldots$. (<u>Fig.3/8</u>)∎

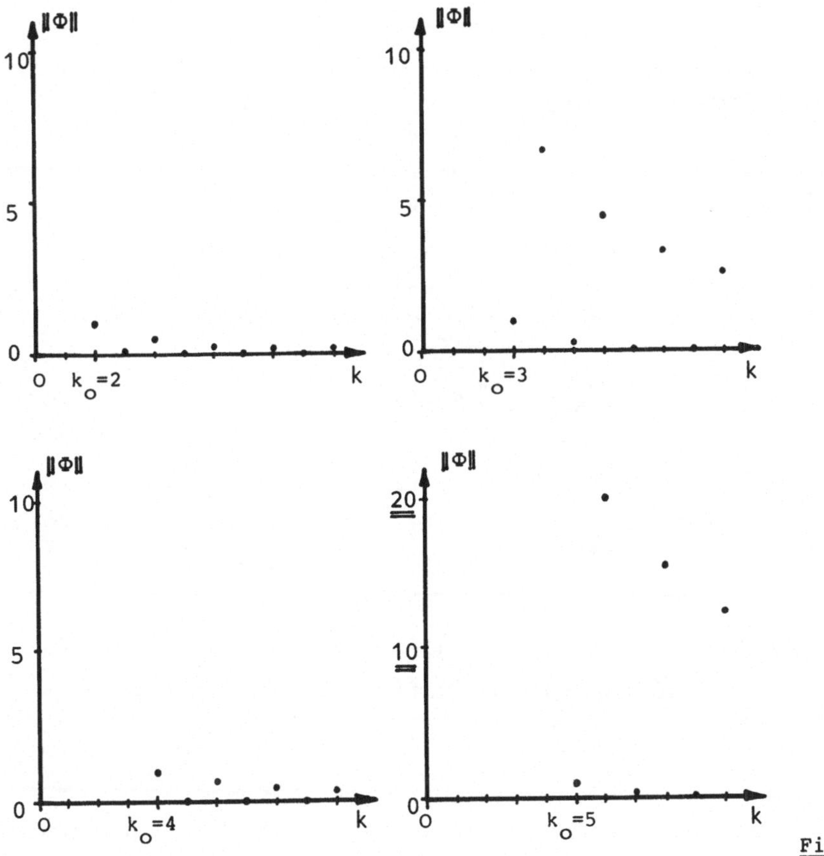

<u>Fig.3/8</u>

The estimations (2.31) and (2.32) yield the sufficient condition:

3-20 | **LEMMA:** If for every $k_o \in K$ and all $k \in [k_o, \omega)$

$$\prod_{i=k_o}^{k-1} \lambda_{Hmax}^{1/2}(i) < \infty \qquad (3.30)$$

resp.

$$\sum_{i=k_o}^{k-1} \ln \lambda_{Hmax}^{1/2}(i) < \infty \qquad (3.31)$$

is valid, then $A(k) \in \mathcal{US}$.

To the class $\mathcal{US}$ corresponds a stronger kind of stability than LJAPUNOW-stability.

3-21 | **DEFINITION:** The solution x of the system $x(k+1) = A(k)x(k)$ is said to be _uniformly stable_, if for every $\varepsilon > 0$ exists a $\delta(\varepsilon) > 0$, such that for every $k_o \in K$ and every $k \in [k_o, \omega)$

$$\| \tilde{x}_o - x_o \| < \delta(\varepsilon) \qquad (3.32)$$

implies

$$\| \tilde{x}(k; k_o, \tilde{x}_o) - x(k; k_o, x_o) \| < \varepsilon . \qquad (3.33)$$

The following lemma can be derived by the same arguments which were needed for the derivation of Lemma 3-5 and Theorem 3-7.

3-22 | **LEMMA:** If a solution of $x(k+1) = A(k)x(k)$ is uniformly stable, then all solutions are uniformly stable.

3-23   THEOREM: The system $x(k+1)=A(k)x(k)$ is uniformly stable if and only if $A(k) \in US$.

For this theorem the following definition was used:

3-24   DEFINITION: The *system* $x(k+1)=A(k)x(k)$ is said to be *uniformly stable*, if every solution is uniformly stable.

If the system matrix A is *time-invariant* and the system $x(k+1) = A x(k)$ is stable, it is uniformly stable too! Because, if $A \in S$, then for $k \geq 0 = \alpha$ there exists a constant $c > 0$ such that

$$\|\Phi(k,0)\| = \|A^k\| < c \ .$$

But since $\Phi(k,k_o) = A^{k-k_o} = \Phi(k-k_o,0)$, it is also for $k \geq k_o$

$$\|\Phi(k,k_o)\| < c \ ,$$

hence $A \in US$.

For time-invariant system matrices A the class $S = US$ can be characterized by the eigenvalues of A and the accompanying eigenvectors:

3-25   THEOREM: The time-invariant linear system $x(k+1)=Ax(k)$ is stable, $A \in S = US$, if and only if

(i) all eigenvalues λ_i of A have absolute value smaller than one or equal to one, $|\lambda_i| \leq 1$, and

(ii) if λ_i is an eigenvalue of absolute value equal to one, and this eigenvalue has multiplicity m, there are exactly m linearly independent eigenvectors.

   This theorem can be proved by help of the JORDAN canonical
form. Because every matrix A can be transformed by a similar-
ity transformation to the form

$$J = T^{-1} A\, T\ ,$$

whereby the matrix J has the form

$$J = \begin{bmatrix} J_1 & & O \\ & \ddots & \\ O & & J_r \end{bmatrix},$$

$$J_i = \begin{bmatrix} J_{i1} & & O \\ & \ddots & \\ O & & J_{in_i} \end{bmatrix},$$

$$J_{ij} = \begin{bmatrix} \lambda_i & 1 & 0 & \cdots & 0 \\ 0 & \lambda_i & 1 & 0 \cdots & 0 \\ \vdots & & \ddots & \ddots & \vdots \\ \vdots & & & \ddots & 1 \\ 0 & \cdots & & 0 & \lambda_i \end{bmatrix},$$

the system matrix A has r distinct eigenvalues, and to the
eigenvalue λ_i, n_i linear independent eigenvectors are
associated. By complete induction one can show, that

$$J_{ij}^k = \begin{bmatrix} \lambda_i^k & \binom{k}{1}\lambda_i^{k-1} & \binom{k}{2}\lambda_i^{k-2} & \cdots & \binom{k}{s-1}\lambda_i^{k-s-1} \\ & \lambda_i^k & \binom{k}{1}\lambda_i^{k-1} & \cdots & \binom{k}{s-2}\lambda_i^{k-s-2} \\ & & \lambda_i^k & & \vdots \\ & & & \ddots & \vdots \\ O & & & & \lambda_i^k \end{bmatrix}. \tag{3.34}$$

For $|\lambda_i| < 1$ the elements of the matrix (3.34) are bounded
for all $k \in [0,\infty)$. If $|\lambda_i| = 1$ the elements of J_{ij}^k are bounded
for $k \to \infty$ only if J_{ij} is a 1×1-matrix. On the other hand

the matrices J_{ij} are 1×1-matrices if and only if there are
exactly m linear independent eigenvalues associated to the
eigenvalue λ_i of multiplicity m.

Condition (ii) of Theorem 3-25 is of course satisfied if
the n×n system matrix A has exactly n distinct eigenvalues.

For *time-variant* systems the conditions of Theorem 3-25
are no longer sufficient, even though the eigenvalues of the
system matrix A(k) are independent of k.

EXAMPLE 3.7: Let K=(0,∞) and the system matrix be

$$
A(k) = \begin{bmatrix} -\cos k\pi/2 & 1+\sin k\pi/2 \\ -1+\sin k\pi/2 & \cos k\pi/2 \end{bmatrix} . \tag{3.35}
$$

Then the characteristic equation is $\det[\lambda I-A(k)]=\lambda^2=0$, and
the eigenvalues are $\lambda_1=\lambda_2=0$. But for the initial state
$x_o^T=[0,1]$ e.g., the norm of the state vector tends to infinite
as k→∞, see Fig.3/9, hence the system $x(k+1)=A(k)x(k)$ is not
stable.∎

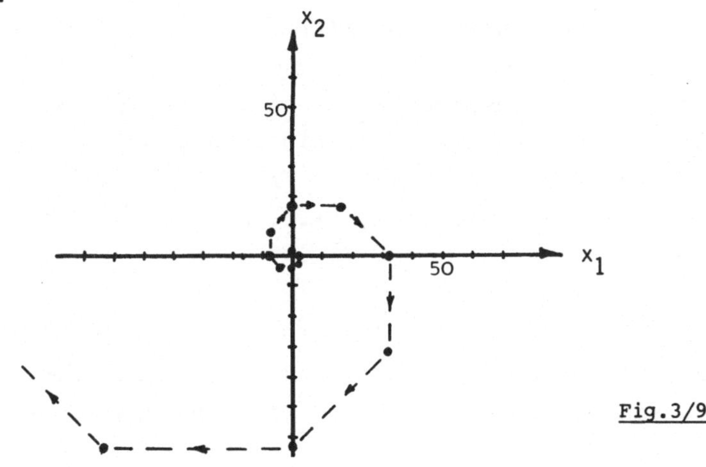

Fig.3/9

For time-variant linear systems the following is rather
true:

<u>3-26</u> **THEOREM:** Let $K=(\alpha,\infty)$. If

  (i) the elements $a_{ij}(k)$ of the system matrix
     $A(k)$ are sequences of bounded variation
     over $[k_2,\infty)$ for a k_2,

  (ii) for the eigenvalues $\lambda_i(k)$ of $A(k)$,
     $|\lambda_i(k)| \leq 1$ is valid for $k \in [k_2,\infty)$, and

  (iii) the eigenvalues of $A_o := \lim_{k \to \infty} A(k)$, whose

     absolute value is equal to one, are
     simple,

  then $A(k) \in \mathbb{US}$.

Before proving this theorem, something has to be said about diagonalization of *time-variant* system matrices A(k). If with the help of the linear transformation

$$x(k) = T(k) \, z(k) \tag{3.36}$$

the system equation $x(k+1)=A(k)x(k)$ shall be transformed into diagonal form, then is

$$z(k+1) = T^{-1}(k+1) \, A(k) \, T(k) \, z(k). \tag{3.37}$$

The time-variant eigenvalues $\lambda_i(k)$ of the system matrix A(k) are obtained from the characteristic equation

$$\det[\lambda(k)I-A(k)] = a_o(k)+a_1(k)\lambda(k)+\ldots+a_{n-1}(k)\lambda^{n-1}(k)+\lambda^n(k)$$

$$= 0 . \tag{3.38}$$

The eigenvalues of the system matrix $T^{-1}(k+1)A(k)T(k)$ in state equation (3.37) one receives from the characteristic equation

$$\det\,[\lambda(k)I - T^{-1}(k+1)A(k)T(k)] = 0. \tag{3.39}$$

The solutions of the equations (3.38) and (3.39) are only

equal if

$$T^{-1}(k+1) \, T(k) = I. \tag{3.40}$$

Because, if (3.40) is valid, then det $T(k+1)$=det $T(k)$, and with (3.39) follows

$$\det \left[\lambda(k)T^{-1}(k+1)T(k) - T(k+1)^{-1}A(k)T(k)\right] =$$

$$= \det T^{-1}(k+1) \, \det \left[\lambda(k)I - A(k)\right] \, \det T(k)$$

$$= \det \left[\lambda(k)I - A(k)\right] = 0.$$

But in general (3.40) is not valid. For this reason the state equation (3.37) is modified by addition and subtraction of the term $T^{-1}(k)A(k)T(k)$ in this way:

$$z(k+1) = T^{-1}(k)A(k)T(k)z(k)+[T^{-1}(k+1)-T^{-1}(k)]A(k)T(k)z(k). \tag{3.41}$$

The goal is to have a system matrix in diagonal form

$$T^{-1}(k)A(k)T(k) = \Lambda(k) = \text{diag}\,\{\lambda_i(k)\}, \tag{3.42}$$

where the $\lambda_i(k)$ are the (distinct) eigenvalues of the system matrix $A(k)$. From (3.42) follows that the transformation matrix $T(k)$ is composed of the eigenvectors of $A(k)$ because (3.42) yields

$$A(k)T(k) = T(k)\Lambda(k),$$

hence

$$A(k)t_i(k) = \lambda_i(k)t_i(k), \tag{3.43}$$

where $t_i(k)$ is the i-th column vector of the transformation matrix $T(k)$. But (3.43) is also the definition equation for the eigenvector $t_i(k)$, who is associated to the eigenvalue $\lambda_i(k)$.

By this transformation the system equation $x(k+1)=A(k)x(k)$ was converted into the form

$$z(k+1) = \Lambda(k)z(k) + [T^{-1}(k+1)-T^{-1}(k)]A(k)T(k)z(k), \qquad (3.44)$$

which agrees so much more with the diagonal form, so much better the difference matrix $T^{-1}(k+1)-T^{-1}(k)$ agrees with the zero matrix.

EXAMPLE 3.8: The time-variant system matrix

$$A(k) = \begin{bmatrix} 1 & a(k) & 0 \\ 0 & 1 & 1 \\ 0 & 12 & -4 \end{bmatrix}$$

has the eigenvalues $\lambda_1=1$, $\lambda_2=2.772...$, $\lambda_3=-5.772...$, and the eigenvectors

$$t_1 = \begin{bmatrix} 1 \\ 0 \\ 0 \end{bmatrix}, \quad t_2 = \begin{bmatrix} a(k)/1.772... \\ 1 \\ 1.772... \end{bmatrix}, \quad t_3 = \begin{bmatrix} -a(k)/6.772... \\ 1 \\ -6.772... \end{bmatrix}.$$

According to (3.44) one gets this quasi-diagonal form

$$z(k+1)=\begin{bmatrix} 1 & 0 & 0 \\ 0 & 2.772... & 0 \\ 0 & 0 & -5.772... \end{bmatrix} z(k)+[a(k+1)-a(k)]\begin{bmatrix} 0 & -1.56... & -0.85... \\ 0 & 0 & 0 \\ 0 & 0 & 0 \end{bmatrix}z(k)$$

Indeed the claim (3.40) is only valid if $a(k+1)=a(k)=$constant for all $k\in K$, since it is

$$T^{-1}(k+1)\ T(k) =$$

$$=\begin{bmatrix} 1 & -.416...a(k+1) & -.083...a(k+1) \\ 0 & .792... & .117... \\ 0 & .207... & -.117... \end{bmatrix}\begin{bmatrix} 1 & .564...a(k) & -.147...a(k) \\ 0 & 1 & 1 \\ 0 & .564... & -.147... \end{bmatrix}$$

$$= I + [a(k+1)-a(k)] \begin{bmatrix} 0 & -.564\ldots & .147\ldots \\ 0 & 0 & 0 \\ 0 & 0 & 0 \end{bmatrix} . \blacksquare$$

Now back to Theorem 3-26, which can also be stated in the following form:

<u>3-27</u> $\quad$ THEOREM: *If* (i) $\quad A_o \in \mathcal{S}$, and

(ii) $\quad$ for the time-variant matrix $A_1(k)$

$$\lim_{k \to \infty} A_1(k) = 0$$

and

$$\sum_{i=k_o}^{\infty} \| A_1(i+1) - A_1(i) \| < \infty \qquad (3.45)$$

is valid, i.e., the matrix $A_1(k)$ is of bounded variation, and

(iii) for the time-variant matrix $A_2(k)$

$$\sum_{i=k_o}^{\infty} \| A_2(k) \| < \infty \qquad (3.46)$$

is valid, and

(iv) $\quad$ for the eigenvalues $\lambda_i(k)$ of the time-variant matrix $[A_o + A_1(k)]$

$$\left| \lambda_i(k) \right| \leq 1$$

is valid for all $k \in [k_o, \infty)$ and

$$\lim_{k \to \infty} \lambda_i(k) = \lambda_i ,$$

where the λ_i's are the eigenvalues of A_o,

then is $A(k) := A_o + A_1(k) + A_2(k) \in \mathcal{US}$.

P_{ROOF}: The transformation (3.36) applied to equation

$$x(k+1) = [A_o + A_1(k) + A_2(k)] \, x(k)$$

yields with the modifications according to (3.41),

$$z(k+1) = \{T^{-1}(k)[A_o+A_1(k)]T(k)+[T^{-1}(k+1)-T^{-1}(k)][A_o+A_1(k)]T(k)+$$
$$+T^{-1}(k+1)A_2(k)T(k)\} \, z(k) \qquad (3.47)$$

$$= \{\Lambda(k) + R(k)\} \, z(k) , \qquad (3.48)$$

where

$$\Lambda(k) := T^{-1}(k)[A_o+A_1(k)]T(k)$$

and

$$R(k) := [T^{-1}(k+1)-T^{-1}(k)][A_o+A_1(k)]T(k)+T^{-1}(k+1)A_2(k)T(k).$$
$$(3.49)$$

Since by assumption (ii) $A_1(k) \to O$ for $k \to \infty$, there must exist a sufficient large k_1, such that for $k \in [k_1, \infty)$ all eigenvalues $\lambda_i(k)$ of the matrix $[A_o+A_1(k)]$ are simple. Therefore a regular (bounded) transformation matrix $T(k)$ one obtains, for which

$$T^{-1}(k)[A_o+A_1(k)]T(k) = \Lambda(k) \qquad (3.50)$$

and det $T(k) \to$ det T for $k \to \infty$, where

$$T^{-1}A_o T = \Lambda = \text{diag} \{\lambda_i\}. \qquad (3.51)$$

For the computation of the transformation matrix $T(k)$ only elementary operations are needed with the elements of the matrices A_o and $A_1(k)$, and with the eigenvalues $\lambda_i(k)$ of the matrix $[A_o+A_1(k)]$. Since all these sequences according to assumption (iii) are bounded and of bounded variation, the same is valid for the matrices

T(k) for all $k \in [k_1, \infty)$. So it is

$$\det T(k) < \infty , \tag{3.52}$$

$$\det T^{-1}(k) < \infty , \tag{3.53}$$

and

$$\sum_{i=k_1}^{\infty} \| T(k+1) - T(k) \| < \infty , \tag{3.54}$$

i.e., the transformation matrix $T(k)$ is of bounded variation too. Now (3.47),(3.48) resp. can be estimated. The i-th row of the vector equation (3.48) has the form

$$z(k+1) = \lambda_i(k) z_i(k) + f_i(k) \; ; \; i \in [1,n], \tag{3.55}$$

where

$$f_i(k) = r_i^T(k) z(k) \tag{3.56}$$

and $r_i^T(k)$ is the i-th row of the matrix $R(k)$. With $c_i := z_i(k_1)$ equation (3.55) has the solution

$$z_i(k) = c_i \prod_{j=k_1}^{k-1} \lambda_i(j) + \sum_{j=k_1}^{k-1} \left[\prod_{s=j+1}^{k-1} \lambda_i(s) \right] f_i(j) \; ; \; i \in [1,n]. \tag{3.57}$$

With $|\lambda_i(k)| \leq 1$ of assumption (i) one obtains for (3.57) the estimation

$$|z_i(k)| \leq |c_i| + \sum_{j=k_1}^{k-1} |r_i^T(j) z(j)| \; ; \; i \in [1,n], \tag{3.58}$$

and therefore with $c := c_1 + \ldots + c_n$

$$\|z(k)\| \leq c + \sum_{j=k_1}^{k-1} \|R(j)\| \|z(j)\| . \tag{3.59}$$

By the assumptions of the theorem and by the estimation
(3.54) follows

$$\sum_{j=k_1}^{\infty} \| R(j) \| < \infty , \tag{3.60}$$

hence one obtains for (3.59) in consequence of Lemma
2-3 the estimation

$$\| z(k) \| \le c \cdot \exp\left[\sum_{j=k_1}^{k-1} \| R(j) \| \right] < \infty , \tag{3.61}$$

thus $\| z(k) \|$ is bounded, i.e., $\| x(k) \|$ is bounded for
$k \in [k_1, \infty)$ too. ∎

But there are systems for which not all assumptions of
Theorem 3-27 are valid although they are stable.

EXAMPLE 3.9: For the time-variant linear system $x(k+1)=A(k)x(k)$
with the system matrix

$$A(k) = \begin{bmatrix} -\frac{1}{2} \cos k\pi/2 & \frac{1}{4} + \frac{1}{2} \sin k\pi/2 \\ -\frac{1}{4} + \frac{1}{2} \sin k\pi/2 & \frac{1}{2} \cos k\pi/2 \end{bmatrix} \tag{3.62}$$

$$= A_0 + A_1(k)$$

$$= \begin{bmatrix} 0 & 1/4 \\ -1/4 & 0 \end{bmatrix} + \begin{bmatrix} -\frac{1}{2} \cos k\pi/2 & \frac{1}{2} \sin k\pi/2 \\ \frac{1}{2} \sin k\pi/2 & \frac{1}{2} \cos k\pi/2 \end{bmatrix} ,$$

assumption (i) of Theorem 3-26, resp. assumption (ii) of
Theorem 3-27 about the bounded variations of the elements of
the matrix $A(k)$, resp. $A_1(k)$ are not valid. But by application
of the following theorem it can be shown, that the system is
uniformly stable! ∎

<u>3-28</u>

> **THEOREM:** If (i) for every $k_o \in K$, there exists a
> $k_1 \in [k_o, \omega)$ such that for all $k \in [k_1, \omega)$
>
> $$\lambda_{Hmax}(k) \leq 1 \qquad\qquad (3.63)$$
>
> is valid, whereby λ_{Hmax} is the largest
> eigenvalue of the hermitian matrix
> $H(k) = A^*(k)A(k)$, and
> (ii) for every $k \in [k_o, k_1)$ $\lambda_{Hmax}(k)$ is bounded
> then $A(k) \in \mathcal{US}$.

PROOF: According to inequality (2.31) the following estimation
for the state transition matrix is valid

$$\|\Phi(k, k_o)\| \leq \prod_{i=k_o}^{k-1} \lambda_{Hmax}^{1/2}(i) \; ,$$

hence for sufficiently large $k \in (k_1, \omega)$

$$\|\Phi(k, k_o)\| \leq \prod_{j=k_o}^{k_1-1} \lambda_{Hmax}^{1/2}(j) \cdot \prod_{i=k_1}^{k-1} \lambda_{Hmax}^{1/2}(i)$$

$$\leq c \prod_{i=k_1}^{k-1} \lambda_{Hmax}^{1/2}(i) \; ,$$

where

$$c := \prod_{j=k_o}^{k_1-1} \lambda_{Hmax}^{1/2}(j) \; .$$

But by assumption (3.63) for all $k \in (k_1, \omega)$

$$\prod_{i=k_1}^{k-1} \lambda_{Hmax}^{1/2}(i) \leq 1$$

is valid, thus

$$\| \Phi(k,k_o) \| \leq c < \infty$$

and therefore, in consequence of Definition 3-19, $A(k) \in \mathcal{US}.\blacksquare$

EXAMPLE 3.10: Now the stability of the time-variant linear system of Example 3.9 with the system matrix (3.62) is checked by the help of Theorem 3-28. For this example the hermitian matrix H(k) has the form

$$H(k) = A^T(k)\, A(k) = \begin{bmatrix} \frac{5}{16} - \frac{1}{4} \sin k\pi/2 & -\frac{1}{4} \cos k\pi/2 \\[2mm] -\frac{1}{4} \cos k\pi/2 & \frac{5}{16} + \frac{1}{4} \sin k\pi/2 \end{bmatrix}$$

and the characteristic equation

$$\det[\lambda_H I - H(k)] = \lambda_H^2 - \frac{5}{8} \lambda_H + \frac{3}{16^2} = 0$$

of the time-variant matrix H(k) in this example is time-invariant. The eigenvalues are

$$\lambda_{H,1} = \lambda_{Hmax} = 9/16 \quad \text{and} \quad \lambda_{H,2} = 1/16.$$

Since for all k condition (3.63) is valid, it follows that $A(k) \in \mathcal{US}$, i.e., the given system is uniformly stable.$\blacksquare$

Now so-called *nearly time-invariant* linear systems are analysed. These are linear time-variant systems with the mathematical description

$$x(k+1) = [A_o + A_1(k)]\, x(k) , \tag{3.64}$$

whereby

$$\lim_{k \to \omega} [A_o + A_1(k)] = A_o , \tag{3.65}$$

i.e., they behave for $k \to \omega$ like time-invariant systems.

<u>3-29</u>

> **THEOREM:** If (i) $A_o \in \mathcal{S}$ (A_o time-invariant), and
>
> $$\text{(ii)} \quad \sum_{i=k_o}^{\omega} \|A_1(i)\| < \infty \qquad (3.66)$$
>
> is valid for all $k_o \in K$,
> then $[A_o + A_1(k)] \in \mathcal{S}$.

PROOF: If in

$$x(k+1) = [A_o + A_1(k)] \, x(k)$$

$$= A_o x(k) + A_1(k) x(k) \qquad (3.67)$$

the term $A_1(k)x(k)$ is interpreted as a forcing sequence $f(.)$, according to (2.8) the solution of (3.67) for $k_o = 0$ is

$$x(k) = A_o^k x_o + \sum_{i=0}^{k-1} A_o^{k-1-i} A_1(i) x(i). \qquad (3.68)$$

In consequence of assumption (i) $\|A_o^k x_o\| < c_1$ and $\|A_o^k\| < c_2$ is valid for $k \geq 0$, i.e., for equation (3.68) one has the estimation with Lemma 2-3

$$\|x(k)\| \leq c_1 \exp\left[\sum_{i=0}^{k-1} c_2 \|A_1(i)\| \right]$$

$$\leq c_1 \exp\left[c_2 \sum_{i=0}^{\omega} \|A_1(i)\| \right] < \infty \,,$$

i.e., $x(k)$ is bounded and therefore the system (3.67) is stable. ∎

A similar theorem is valid for *time-variant* systems.

<u>3-30</u>

> THEOREM: If (i) $A_o(k) \in \mathcal{S}$, and
>
> (ii) for $A_1(k)$ the bound (3.66) is valid,
>
> then $[A_o(k) + A_1(k)] \in \mathcal{S}$.   (3.69)

PROOF: With

$$x(k+1) = A_o(k)x(k) + A_1(k)x(k)$$

and $f(k):=A_1(k)x(k)$ the proof is the same as the proof for Theorem 3-29.∎

If the system matrix A(k) of an uniformly stable system is regular, the system matrix can be transformed into a very simple form. First the following definition:

<u>3-31</u>

> DEFINITION: The system matrix A(k) of the time-variant
> linear system x(k+1)=A(k)x(k) is said to
> be <u>*reducible*</u>, if there exist regular trans-
> formation matrices T(k) for all k ∈ K,
> such that $T^{-1}(k+1)A(k)T(k)$ is a time-in-
> variant matrix for k ∈ K. If a transfor-
> mation
>
> $$T^{-1}(k+1)A(k)T(k) = I \qquad (3.70)$$
>
> is possible for k ∈ K, the system matrix
> A(k) is said to be <u>*reducible to one*</u>.

Now the next theorem can be stated.

<u>3-32</u>

> THEOREM: The linear time-variant system x(k+1)=A(k)x(k)
> with a regular system matrix A(k) for k ∈ K
> is uniformly stable if and only if it is
> reducible to one.

Proof: If the linear time-variant system is uniformly stable,
then $\|\Phi(k,k_o)\| < c < \infty$ for every $k \geq k_o, k \in K$ is valid.
Since $A(k)$ is assumed to be regular for $k \in K$, $\Phi(k,k_o) =$
$A(k-1)...A(k_o)$ is regular too. Therefore the transition
matrix $\Phi(k,k_o)$ is invertible and equation (2.12) can
be converted by multiplication from the left with
$\Phi^{-1}(k+1,k_o)$ into

$$\Phi^{-1}(k+1,k_o)A(k)\Phi(k,k_o) = I \ ,$$

hence the system matrix $A(k)$ can be reduced to one
with the help of the transformation matrix $T(k) = \Phi(k,k_o)$.
Contrary, if $T(k)$ is a matrix which is able to reduce
$A(k)$ to one, then $T^{-1}(k+1)A(k)T(k)=I$ is valid, i.e.,
$T(k+1)=A(k)T(k)$, hence according to equation (2.12),
$T(k)$ is the transition matrix, and because the inverse
exists, the transition matrix must be regular and
bounded. Thus the system must be uniformly stable.∎

The assumed regularity of the system matrix $A(k)$ in
Theorem 3-32 indeed limits the validity of the theorem. But e.g.
for the large system class of sampled-data-systems the system
matrix is always regular.

3.4 Asymptotic Stability

In case of LJAPUNOW-stability there was only required that
a new motion $\tilde{x}$ remains in the neighbourhood of the old motion
x, if the initial state was changed only minimal. But e.g. in
technical systems is needed more, namely that the new motion
$\tilde{x}$ tends for increasing time to the old motion x.

3-33 DEFINITION: The solution x of $x(k+1)=A(k)x(k)$ is said
to be *asymptotically stable*, if the
solution is LJAPUNOW-stable and additional

$$\lim_{k\to\omega}\|\tilde{x}(k;k_o,\tilde{x}_o) - x(k;k_o,x_o)\| = 0 \qquad (3.71)$$

is valid. (<u>Fig 3/10</u>)

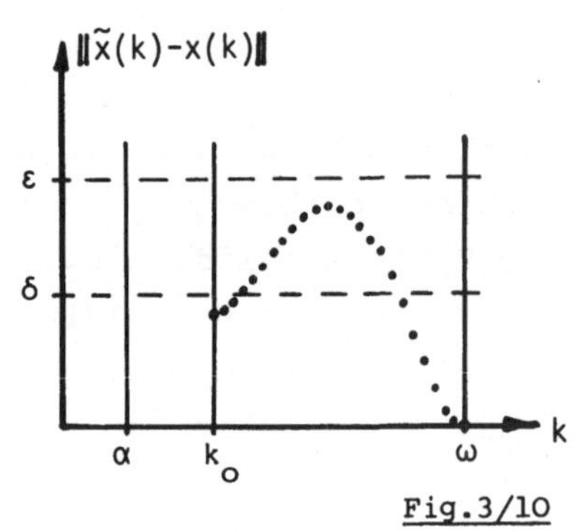

Fig.3/10

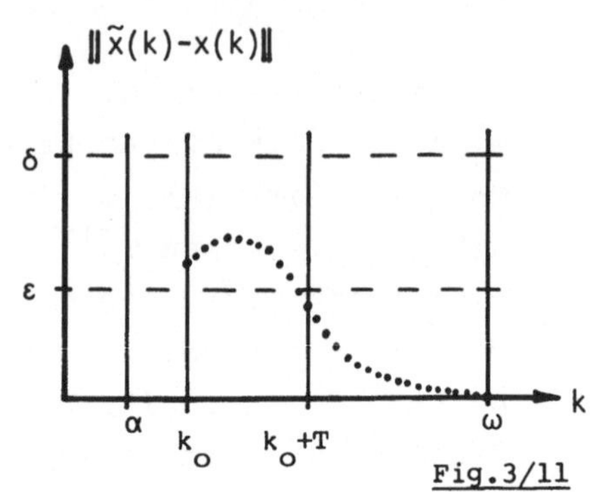

Fig.3/11

An equivalent definition of asymptotic stability is given
in the following definition (see <u>Fig.3/11</u>).

3-34 DEFINITION: The solution x of $x(k+1)=A(k)x(k)$ is said
to be *asymptotically stable*, if the
solution is LJAPUNOW-stable and for every
$k_o \in K$ and every $\varepsilon > 0$ exists a $\delta(k_o) > 0$
and a $T = T(\varepsilon,k_o) \geq 0$ such that
$\|\tilde{x}_o - x_o\| < \delta(k_o)$ implies

$$\|\tilde{x}(k;k_o,\tilde{x}_o) - x(k;k_o,x_o)\| < \varepsilon \qquad (3.72)$$

for all $k \in [k_o+T,\omega)$.

Since the solution of $x(k+1)=A(k)x(k)$ is $x(k)=\Phi(k,k_o)x_o$, condition (3.71) is equivalent to condition (3.75) of the following definition:

3-35   DEFINITION: The matrix $A(k)$ is said to be of *class* **A$**,
$A(k) \in$ **A$**, if $A(k) \in$ **$** and additional

$$\lim_{k\to\omega} \Phi(k,k_o) = 0 \qquad\qquad (3.73)$$

is valid for every $k_o \in K$.

For *continuous-time* time-variant linear systems $x(t)=F(t)x(t)$ the conditions (3.71), resp. (3.73) can never be valid for a *finite* time interval and an initial state $x_o \neq 0$, because the state transition matrix $\Phi_F(t,t_o)$ is never singular! But in case of *discrete-time* time-variant linear systems it is possible!

EXAMPLE 3.11: For the discrete-time time-variant linear system

$$x(k+1) = \begin{bmatrix} -2\ a(k) & -4\ a(k) \\ a(k) & 2\ a(k) \end{bmatrix} x(k)$$

is

$$\Phi(k+2,k) = \begin{bmatrix} -2a(k+1) & -4a(k+1) \\ a(k+1) & 2a(k+1) \end{bmatrix} \cdot \begin{bmatrix} -2a(k) & -4a(k) \\ a(k) & 2a(k) \end{bmatrix} = 0$$

for every k, even if $a(k) \neq 0$ for every k. ∎

EXAMPLE 3.12: Another example for finite time asymptotically stable solutions is the class of discrete-time time-invariant linear systems $x(k+1)=Ax(k)$, for which the time-invariant system matrix A is *nilpotent*, because for a nilpotent matrix of index q is $A^q = \Phi(q,0) = 0$. $\blacksquare$

The following lemma and theorem can be obtained in the same way as Lemma 3-5 and Theorem 3-7 above.

<table>
<tr><td>3-36</td><td>LEMMA: If a solution of $x(k+1)=A(k)x(k)$ is asymptotically stable, then all solutions are asymptotically stable.</td></tr>
<tr><td>3-37</td><td>THEOREM: The system $x(k+1)=A(k)x(k)$ is asymptotically stable if and only if $A(k) \in AS$.</td></tr>
</table>

For Theorem 3-37 the following definition was used:

<table>
<tr><td>3-38</td><td>DEFINITION: The <u>*system*</u> $x(k+1)=A(k)x(k)$ is said to be <u>*asymptotically stable*</u>, if every solution is asymptotically stable.</td></tr>
</table>

Some conditions for the membership of $A(k)$ to the class AS shall be stated now. From the estimation (2.31)

$$\| \Phi(k,k_o) \| \leq \prod_{i=k_o}^{k-1} \lambda_{Hmax}^{1/2}(i) \tag{3.74}$$

for the norm of the state transition matrix immediately follows the sufficient condition for $A(k) \in AS$:

<table>
<tr><td>3-39</td><td>LEMMA: If for every $k_o \in K$

$$\prod_{i=k_o}^{\omega-1} \lambda_{Hmax}^{1/2}(i) = 0 \tag{3.75}$$

is valid, then $A(k) \in AS$.</td></tr>
</table>

From this lemma the two Lemmata 3-40 and 3-41 are following:

3-40   LEMMA: If for every $k_o \in K$ there exists a $k \in [k_o, \omega)$ such that

$$\lambda_{Hmax}(k) = 0 , \qquad (3.76)$$

then $A(k) \in A\mathcal{S}$.

3-41   LEMMA: Let K be a finite time interval, i.e. $\omega < \infty$. Then $A(k) \in A\mathcal{S}$ if and only if for every $k_o \in K$ and some $k \in [k_o, \omega)$ equation (3.76) is valid.

The condition (3.73) yields further

$$\lim_{k \to \omega} \det \Phi(k, k_o) = 0 , \qquad (3.77)$$

therefore with (2.54), namely

$$\det \Phi(k, k_o) = \prod_{i=k_o}^{k-1} \det A(i) ,$$

the necessary condition:

3-42   LEMMA: If $A(k) \in A\mathcal{S}$, then

$$\lim_{k \to \omega} \prod_{i=k_o}^{k-1} \det A(i) = 0 . \qquad (3.78)$$

As usual in Definition 3-33 of an asymptotically stable solution was beside the convergence of the state vector to the zero vector additionally required the LJAPUNOW-stability of the solution. But this is not necessary for *linear* systems, because for linear systems $x(k+1) = A(k)x(k)$ the following lemma is valid:

3-43 **LEMMA:** If the convergence condition (3.71) is valid,
then the linear system $x(k+1)=A(k)x(k)$ is
LJAPUNOW-stable.

PROOF: If (3.71) is valid, then according to Theorem 3-37 the
condition (3.73) is valid too. Moreover all elements
of the state transition matrix $\Phi(k,k_o)$ have to be
bounded for $k \in K$, since this is a permanent assumption
for the elements of the system matrix $A(k)$. Then for
the matrix

$$\Psi(k,k_o) := \Phi^*(k,k_o)\,\Phi(k,k_o)$$

there has to exist a constant $b < \infty$, such that for all
elements $\psi_{ij}(k,k_o)$ of the matrix $\Psi(k,k_o)$

$$|\psi_{ij}(k,k_o)| \leq b$$

is valid for all $k \in K$. Further one gets the estimation

$$\|x(k;k_o,x_o)\| = [x_o^*\Phi^*(k,k_o)\Phi(k,k_o)x_o]^{1/2}$$

$$= [x_o^*\Psi(k,k_o)x_o]^{1/2}$$

$$\leq \|x_o\|\left[\sum_{i,j=1}^{n} |\psi_{ij}(k,k_o)|^2\right]^{1/4}$$

$$\leq \|x_o\|\,(n\cdot b)^{1/2} \quad ; \quad k \in K.$$

Hence for a given $\varepsilon > 0$ is $\|x(k;k_o,x_o)\| < \varepsilon$ if $\|x_o\| < \gamma$
and if $\|x_o\|(n\cdot b)^{1/2} < \varepsilon$, i.e., if $\|x_o\| < \delta$, where
$\delta := \min\,[\gamma,\varepsilon(nb)^{-1/2}]$. ∎

Lemma 3-43 shows immediately, that the class $A\mathcal{S}$ is a sub-
class of the class $\mathcal{S}$:

$$A\mathcal{S} \subset \mathcal{S}. \tag{3.79}$$

But as Example 3.4 shows, it is $A\mathcal{S} \neq \mathcal{S}$ and $A\mathcal{S} \neq \mathcal{U}\mathcal{S}$!

For the class of *time-invariant* linear systems easily follows from the form of J^k_{ij} in (3.34):

<table>
<tr><td>3-44</td><td>

THEOREM: Let the system matrix A be time-invariant. Then is $A \in A\mathcal{S}$ if and only if for all eigenvalues λ of A

$$|\lambda| < 1 \tag{3.80}$$

is valid.

</td></tr>
</table>

Further the following theorem is valid for *time-variant* systems:

<table>
<tr><td>3-45</td><td>

THEOREM: Let the time interval K be unbounded, i.e. $\omega = \infty$. If a $k_1 \geq k_0$ exists such that for all $k \geq k_1$

$$\lambda_{Hmax}(k) < 1 \tag{3.81}$$

is valid, then $A(k) \in A\mathcal{S}$.

</td></tr>
</table>

PROOF: If (3.81) is valid, then exists a β such that for all $k \in [k_1, \infty)$ one has

$$\lambda_{Hmax}(k) \leq \beta < 1 \, ,$$

thus in consequence of (2.31) the estimation for the state transition matrix

$$\|\Phi(k,k_O)\| \leq \prod_{i=k_O}^{k-1} [+\lambda_{Hmax}^{1/2}(i)]$$

$$= c \prod_{i=k_1}^{k-1} [+\lambda_{Hmax}^{1/2}(i)]$$

$$\leq c \, \beta^{k-1-k_1}.$$

Hence

$$\lim_{k \to \infty} \|\Phi(k,k_O)\| \leq \lim_{k \to \infty} c \, \beta^{k-1-k_1} = 0 \, ,$$

therefore from Definition 3-35 $A(k) \in A\$$ is following.∎

3.5 ℓ^P-STABILITY

The now introduced stability class is needed first of all
for input-state stability and input-output stability in the
next chapters 4 and 5.

In physically and technically systems energy considerations
are very important, where e.g. the energy of a physically
quantity u is given by the integral

$$\left[\int_{-\infty}^{+\infty} |u(t)|^2 dt \right]^{1/2}.$$

Therefore a further subclass of $\$$ which is distinct both from
subclasses $U\$$ and $A\$$ is defined by:

<table>
<tr><td>3-46</td><td>DEFINITION: The Matrix A(k) is said to be of class $\ell^P\$$,
$A(k) \in \ell^P\$$, for any $p \geq 1$, if for every $k_O \in K$
exists a positive constant $c_p(k_O) < \infty$, such</td></tr>
</table>

$$\begin{array}{|l}
\text{that for all } k \in [k_o, \omega) \\[4pt]
\left[\displaystyle\sum_{i=k_o}^{k-1} \| \Phi(k,i) \|^p \right]^{1/p} < c_p(k_o) \qquad\qquad (3.82) \\[6pt]
\text{is valid.}
\end{array}$$

The next lemma shows that the new defined class is a sub-class of $\mathcal{S}$.

$$\underline{3\text{-}47} \quad \boxed{\begin{array}{l}
\textsc{Lemma: For } p \geq 1 \text{ the inclusion} \\[6pt]
\ell^p\mathcal{S} \subset \mathcal{S} \qquad\qquad\qquad\qquad (3.83) \\[6pt]
\text{is valid.}
\end{array}}$$

$\textsc{Proof:}$ If $A(k) \in \ell^p\mathcal{S}$, the estimation (2.53) of the state transition matrix gives for $k \geq k_1 > k_o$

$$\| \Phi(k,k_1) \| \leq c_p(k_o) \left[\sum_{i=k_o}^{k_1-1} \| \Phi(i,k_1) \|^{-p} \right]^{-1/p} \left[\sum_{j=k_1}^{k-1} \left\{ \sum_{i=k_o}^{j-1} \| \Phi(j,i) \|^p \right\}^{-1} \right]^{-1/p}$$

$$\leq c_p(k_o) \left[\sum_{i=k_o}^{k-1} \| \Phi(i,k_1) \|^{-p} \right]^{-1/p} \left[c_p^{-p}(k_o) \ (k-k_1) \right]^{-1/p}$$

$$= c_p^2(k_o) \ (k-k_1)^{-1/p} \left[\sum_{i=k_o}^{k-1} \| \Phi(i,k_1) \|^{-p} \right]^{-1/p} , \qquad (3.84)$$

hence $A(k) \in \ell^p\mathcal{S}$. $\blacksquare$

Since the right hand side of inequality (3.84) tends to zero for $k \to \infty$, one has the following inclusion:

<u>3-48</u>

> **LEMMA:** Let the time interval unbounded, i.e., $\omega=\infty$.
> Under this condition the inclusion
>
> $$\ell^p\mathcal{S} \subset A\mathcal{S} \qquad\qquad (3.85)$$
>
> is valid for $p\geq 1$.

If the time interval K is finite, i.e. if $\omega<\infty$, the inclusion (3.85) is not longer valid. This is shown by the following example.

EXAMPLE 3.13: For $f(k)=1 + 2^{1/k}$ in the first order system $x(k+1)=f(k+1)f^{-1}(k)x(k)$ of Example 3.5 the 'state transition matrix' is

$$\Phi(k,k_o) = f(k)f^{-1}(k_o)$$

$$= (1 + 2^{1/k})(1 + 2^{1/k_o})^{-1}. \qquad (3.86)$$

For $K=(-\infty,0]$ is

$$\lim_{k\to 0} \Phi(k,k_o) = (1 + 2^{1/k_o})^{-1},$$

thus $A(k) \notin A\mathcal{S}$. But since $|\Phi(k,k_o)|<2$ for all $k\in[k_o,0)$ follows that

$$\left[\sum_{i=k_o}^{k-1} |\Phi(k,i)|^p\right]^{1/p} < \left[2^p \sum_{i=k_o}^{k-1} 1\right]^{1/p}$$

$$= 2(k-k_o)^{1/p}$$

$$< 2(-k_o)^{1/p} =: c_p(k_o),$$

i.e., $A(k) \in \ell^p\mathcal{S}$. $\blacksquare$

On the other side the set $A\mathcal{S}\setminus\ell^p\mathcal{S}$ is not empty, i.e., there are system matrices A(k) for which A(k)$\in A\mathcal{S}$ is valid, but it is A(k)$\notin\ell^p\mathcal{S}$. Such a system is shown in the following example.

EXAMPLE 3.14: For $f(k)=k^{-2+\cos k\pi/2}$ in the first order system $x(k+1)=f(k+1)f^{-1}(k)x(k)$ the 'state transition matrix' is

$$\Phi(k,k_o) = k^{-2+\cos k\pi/2}\, k_o^{2-\cos k_o\pi/2}\,,$$

thus for $K=(0,\infty)$ is

$$\lim_{k\to\infty}\Phi(k,k_o) = 0$$

for all $k_o\in K$, i.e., A(k)$\in A\mathcal{S}$. But since $\cos k_o\pi/2\le 1$ for every k_o is valid, one has for $k_o>1$ that $k_o^{2-\cos k_o\pi/2}\ge k_o$; hence for $k\ge k_o\ge 1$ the following estimation is valid

$$\left[\sum_{i=k_o}^{k-1}|\Phi(k,i)|^p\right]^{1/p} \ge k^{-2+\cos k\pi/2}\left[\sum_{i=k_o}^{k-1}i^p\right]^{1/p}. \tag{3.87}$$

The inequality

$$\sum_{i=1}^{k}i^p \ge \frac{k^{p+1}}{p+1} \ge \sum_{i=1}^{k-1}i^p \tag{3.88}$$

can be proved by mathematical induction. The following inequality can be deduced from (3.88)

$$\sum_{i=k_o}^{k-1}i^p \ge \frac{(k-1)^{p+1}-k_o^{p+1}}{p+1}. \tag{3.89}$$

Inequality (3.89) in (3.87) yields

$$\left[\sum_{i=k_o}^{k-1}|\Phi(k,i)|^p\right]^{1/p} \ge k^{-2+\cos k\pi/2}\left[\frac{(k-1)^{p+1}-k_o^{p+1}}{p+1}\right]^{1/p}$$

$$= \frac{k^{-2+\cos k\pi/2}(k-1)^{1+1/p}}{(p+1)^{1/p}} \left[1 - \left(\frac{k_o}{k-1}\right)^{p+1}\right]^{1/p}$$

and finally especially for k=4i (i=1,2,3,...)

$$\sum_{j=k_o}^{4i-1} \left|\Phi(4i,j)\right|^p \geq (1-\tfrac{1}{4i})^{p+1}\frac{4i}{p+1}\left[1 - \left(\frac{k_o}{4i-1}\right)^{p+1}\right].$$

Since the right hand side exceeds all bounds for i→∞, it follows A(k) $\notin$ $\ell^p \mathcal{S}$.∎

It shall be shown with the next example, that systems exist for which A(k) $\in$ $\mathbf{U}\mathcal{S}$ ∩ $\mathbf{A}\mathcal{S}$, but A(k) $\notin$ $\ell^p \mathcal{S}$, i.e., the set ($\mathbf{U}\mathcal{S}$ ∩ $\mathbf{A}\mathcal{S}$)$\smallsetminus\ell^p \mathcal{S}$ does not be empty.

EXAMPLE 3.15: Let K=(0,∞) and f(k)=k^{-1} in the first order system of Example 3.13. Then the 'state transition matrix' is $\Phi(k,k_o)=k_o/k$, thus A(k) $\in$ $\mathbf{U}\mathcal{S}$ ∩ $\mathbf{A}\mathcal{S}$. But with (3.89) it is

$$\sum_{i=k_o}^{k-1} \left|\Phi(k,i)\right|^p = \sum_{i=k_o}^{k-1} (\tfrac{i}{k})^p$$

$$\geq \frac{k^{-p}}{p+1}\left[(k-1)^{p+1} - k_o^{p+1}\right]$$

$$= \frac{k}{p+1}(1-\tfrac{1}{k})^{p+1}\left[1 - (\tfrac{k_o}{k-1})^{p+1}\right]$$

thus

$$\lim_{k\to\infty} \sum_{i=k_o}^{k-1} \left|\Phi(k,i)\right|^p = \infty$$

for all k$_o$∈K, i.e., A(k) $\notin$ $\ell^p \mathcal{S}$.∎

If the system is uniformly stable, A(k) $\in$ $\mathbf{U}\mathcal{S}$, one has in consequence of Definition 3-19

$$\|\Phi(k,k_1)\| < c$$

and for $k \geq k_0$

$$\sum_{i=k_0}^{k-1} \|\Phi(k,i)\|^p < c^p(k-k_0),$$

thus it is $A(k) \in \ell^p\mathcal{S}$ if $\omega < \infty$, i.e., there exist systems for which

$$A(k) \in \ell^p\mathcal{S} \cap \mathcal{US} \tag{3.90}$$

is valid.

With $\mathcal{S} = \mathcal{US}$ and Lemma 3-47 follows for *time-invariant* systems:

3-49 | LEMMA: If the system $x(k+1) = Ax(k)$ is time-invariant, $\ell^p\mathcal{S} \subset \mathcal{US}$ is valid for $p \geq 1$.

3.6 EXPONENTIAL AND UNIFORM ASYMPTOTIC STABILITY

A further stability class is introduced by the next definitions for the infinite time interval $K = (\alpha, \infty)$.

3-50 | DEFINITION: The matrix $A(k)$ is said to be of *class $\mathcal{ES}$*, $A(k) \in \mathcal{ES}$, if there exist positive constants $c < \infty$ and $\beta \in [0,1)$ such that for every $k_0 \in K = (\alpha, \infty)$ and all $k \in [k_0, \infty)$

$$\|\Phi(k,k_0)\| \leq c\, \beta^{\,k-k_0} \tag{3.91}$$

is valid.

DEFINITION: The solution x of $x(k+1)=A(k)x(k)$ is said to be *exponentially stable*, if there exist postive constants $c < \infty$ and $\beta \in [0,1)$ such that for every $k_o \in K$ and all $k \in [k_o, \infty)$

$$\|x(k;k_o,x_o)\| \leq c\,\beta^{k-k_o}\|x_o\| \qquad (3.92)$$

is valid. The *system* $x(k+1)=A(k)x(k)$ is said to be *exponentially stable*, if every solution is exponentially stable. (Fig.3/12)

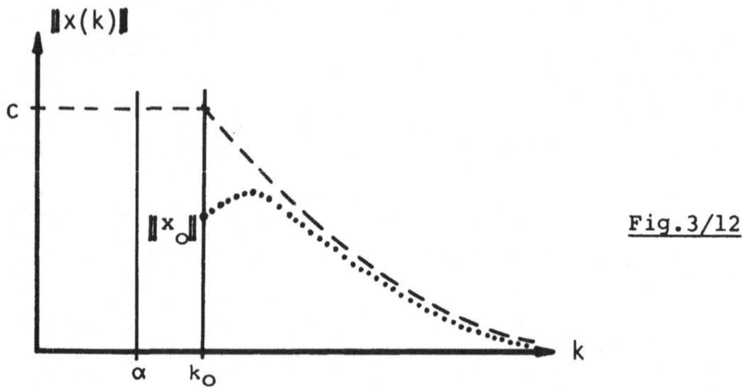

Fig.3/12

From this two definitions immediately Theorem 3-52 is following:

3-52

THEOREM: The system $x(k+1)=A(k)x(k)$ is exponentially stable, if and only if $A(k) \in ES$.

An important relation is written down in the following lemma.

3-53

LEMMA: For $p \geq 1$ the relation

$$ES = \ell^p S \cap US \qquad (3.93)$$

is valid.

PROOF: The relation $E\mathcal{S} \subset \ell^p\mathcal{S} \cap \mathcal{US}$ is following directly from the Definitions 3-19, 3-46, and 3-50. Now the relation $\ell^p\mathcal{S} \cap \mathcal{US} \subset E\mathcal{S}$ shall be verified. Let $A(k) \in \ell^p\mathcal{S} \cap \mathcal{US}$. From the estimation (2.52) for the state transition matrix follows for $k_o \leq k_1 \leq k$:

$$\|\Phi(k,k_1)\| \leq \left[\sum_{i=k_o}^{k-1} \|\Phi(k,i)\|^p\right]^{1/p} \left[\sum_{i=k_o}^{k-1} \|\Phi(i,k_1)\|^{-p}\right]^{-1/p}$$

$$\leq c_p(k_o)\, c\, (k-k_o)^{-1/p}. \tag{3.94}$$

With

$$T := 2^p\, c_p^p(k_o)\, c^p$$

i.e.,

$$c_p(k_o)\, c = \frac{1}{2}\, T^{1/p}, \tag{3.95}$$

for $k > T + k_1 > k_1 \geq k_o$ one has from (3.94) the estimation

$$\|\Phi(k,k_1)\| \leq \frac{1}{2}.$$

Since the above defined T does not depend on k_1, the following estimation is valid for $k > k_1 + (i-1)T > k_1 \geq k_o$:

$$\|\Phi(k,k_1+iT)\| \leq \frac{1}{2} \; ; \; i=1, 2, 3,\ldots,$$

thus for $k > k_1 + jT$ $(j=1, 2, 3,\ldots)$:

$$\|\Phi(k,k_1)\| \leq \|\Phi(k,k_1+(j-1)T)\|\cdot\|\Phi(k_1+(j-1)T,k_1+(j-2)T)\|\cdots\|\Phi(k_1+T,k_1)\|$$

$$\leq \left(\frac{1}{2}\right)^j.$$

Since also $A(k) \in \mathcal{US}$ one has for $k \geq k_1 \geq k_o$:

$$\|\Phi(k,k_1)\| < c,$$

where $c \geq 1$, thus for $k \geq k_1 + jT$ the estimation

$$\| \Phi(k, k_1) \| < c \, 2^{-j}$$

is valid. But if $k > k_1$, then $k_1 + jT \leq k < k_1 + (1+j)T$ for some $j = 0, 1, 2, \ldots$, i.e. for $k \geq k_1 \geq k$ the following estimation is valid:

$$\| \Phi(k, k_1) \| \leq c \, 2^{-j}$$
$$= 2 \, c \, 2^{-(j+1)}$$
$$= 2 \, c \, 2^{-(j+1)T/T}$$
$$\leq 2 \, c \, 2^{-(k-k_1)/T}$$
$$= 2 \, c \left[2^{-1/T} \right]^{k-k_1}.$$

With $c' := 2c$ and $\beta := 2^{-1/T}$ one has $A(k) \in E\$.\blacksquare$

Lemma 3-53 immediately yields $E\$ \subset \ell^p\$$ for all $p \geq 1$. With this relation the following theorem can be derived.

<table>
<tr><td>3-54</td><td>

THEOREM: If $A(k) \in E\$$, then there exist positive constants $c_1 < \infty$ and $c_2 < \infty$ such that for all $k_o \in K$, $k \in [k_o, \infty)$ the following two estimations are valid:

$$\text{(i)} \quad \sum_{i=k_o}^{k-1} \| \Phi(k, i) \| < c_1, \text{ and} \qquad (3.96)$$

$$\text{(ii)} \quad \sum_{i=k_o}^{k-1} \| \Phi(k, i) \|^2 < c_2. \qquad (3.97)$$

</td></tr>
</table>

P_{ROOF}: (i) If $A(k) \in \pmb{E}\pmb{S}$, then is $A(k) \in \ell^1\pmb{S}$ too, i.e., for the state transition matrix Φ as a consequence of Definition 3-46 the inequality (3.96) is valid.

(ii) If $A(k) \in \pmb{E}\pmb{S}$, then $A(k) \in \ell^2\pmb{S}$. Hence

$$\left[\sum_{i=k_o}^{k-1} \|\Phi(k,i)\|^2 \right] < c$$

and therefore

$$\sum_{i=k_o}^{k-1} \|\Phi(k,i)\|^2 < c^2 =: c_2 . \blacksquare$$

In stability theory it is usual to introduce additional the following kind of stability and stability class.

<table>
<tr><td>3-55</td><td>

$\text{D}_{\text{EFINITION}}$: The solution x of $x(k+1)=A(k)x(k)$ is said to be <u>*uniformly asymptotically stable*</u>, if the solution is uniformly stable and additional for all $k_o \in K$

$$\lim_{k\to\infty} \|\tilde{x}(k;\ k_o,\tilde{x}_o) - x(k;k_o,x_o)\| = 0 \qquad (3.98)$$

is valid. The <u>*system*</u> $x(k+1)=A(k)x(k)$ is said to be <u>*uniformly asymptotically stable*</u>, if every solution is uniformly asymptotically stable.

</td></tr>
</table>

<table>
<tr><td>3-56</td><td>

$\text{D}_{\text{EFINITION}}$: The matrix $A(k)$ is said to be of <u>*class* $\pmb{U}\pmb{A}\pmb{S}$</u>, $A(k) \in \pmb{U}\pmb{A}\pmb{S}$, if

(i) $A(k) \in \pmb{U}\pmb{S}$, and

(ii) for every $k_o \in K=(\alpha,\infty)$

$$\lim_{k\to\infty} \Phi(k,k_o) = 0 \qquad (3.99)$$

is valid.

</td></tr>
</table>

An equivalent definition of uniform asymptotic stability
is given in:

<u>3-57</u>

> DEFINITION: The solution x of $x(k+1)=A(k)x(k)$ is said
> to be *uniformly asymptotically stable*, if
> the solution is uniformly stable and for
> every $\varepsilon>0$ there exists a $\delta(\varepsilon)>0$ and a
> $T(\varepsilon)\geq 0$, such that for all $k_o \in K$ and
> $k \in [k_o+T,\infty)$ the implication is valid:
>
> $$\|\tilde{x}_o-x_o\|<\delta \Rightarrow \|\tilde{x}(k;k_o,\tilde{x}_o)-x(k;k_o,x_o)\|<\varepsilon. \qquad (3.100)$$

But of course for *linear* systems the following lemma is
valid:

<u>3-58</u>

> LEMMA: The relation
>
> $$\mathbf{ES} = \mathbf{UAS} \qquad (3.101)$$
>
> is valid, i.e., a linear system $x(k)=A(k)x(k)$
> is exponentially stable, if and only if it is
> uniformly asymptotically stable.

PROOF: (i) $\mathbf{ES} \subset \mathbf{UAS}$. If (3.91) is valid for $\beta \in [0,1]$, then
obviously condition (3.99) is valid too.
(ii) $\mathbf{UAS} \subset \mathbf{ES}$. Since condition (3.99) is valid uniformly with respect to k_o, there exists an integer
$T>0$ such that for all $k_o \in K$

$$\|\Phi(k_o+T,k_o)\| \leq \frac{1}{2} \qquad (3.102)$$

is valid. Furthermore with (2.11) it is

$$\Phi(k_o+iT,k_o)=\Phi(k_o+iT,k_o+(i-1)T)\Phi(k_o+(i-1)T,k_o+(i-2)T)\cdots\Phi(k_o+T,k_o),$$

and with the help of (3.102) one gets the estimation

$$\|\Phi(k_o+iT,k_o)\| \leq \|\Phi(k_o+iT,k_o+(i-1)T)\|\|\Phi(k_o+(i-1)T,k_o+(i-2)T)\|$$

$$\cdots \|\Phi(k_o+T,k_o)\|$$

$$\leq \left(\tfrac{1}{2}\right)^i$$

$$= \left[2^{-1/T}\right]^{iT}. \tag{3.103}$$

With $c:=1$, $\beta:=2^{-1/T}$, $k_1=k_o$, and $k=k_o+iT$ estimation
·(3.103) yields estimation (3.91), i.e., $A(k) \in E\$$.
Now from (i) and (ii) one obtains relation (3.101).■

In the case of *time-invariant* systems $x(k+1)=Ax(k)$ and an
infinite time interval $k=(\alpha,\infty)$ one obtains immediately from
Definition 3-35 and Definition 3-50 the relation $A\$ = E\$$.
Since according to Lemma 3-49 for time-invariant systems
$\ell^P\$ \subset U\$$ is valid, one obtains from Lemma 3-53 and Lemma 3-58:

<table>
<tr><td>3-59</td><td>

LEMMA: For time-invariant systems $x(k+1)=Ax(k)$, infinite
time interval $K=(\alpha,\infty)$, and $p\geq 1$

$$A\$ = E\$ = UA\$ = \ell^P\$ \tag{3.104}$$

is valid.

</td></tr>
</table>

In Example 3.15 the system matrix is a member of the
classes $A\$$ and $U\$$, thus $A(k) \in A\$ \cap U\$$. But it is not a member
of the class $E\$$, since for the state transition matrix
$\Phi(k,k_o) = k_o/k$ there exists no β, such that condition (3.91)
is valid. Therefore it is

$$(U\$ \cap A\$)\setminus E\$ \neq \emptyset. \tag{3.105}$$

Finally a criterion for the exponential stability of

somewhat disturbed systems is given.

3-60

THEOREM: Let $A(k) \in \mathbf{ES}$, i.e., there exist positive constants $c < \infty$ and $\beta \in [0,1)$ such that

$$\|\Phi(k,k_o)\| \leq c \, \beta^{k-k_o}$$

for $k \geq k_o$. If $\|A'(k)\| \leq \delta$ for all $k \in K$, then for the state transition matrix $\Psi(k,k_o)$ of the linear system

$$x(k+1) = [A(k)+A'(k)] \, x(k) \tag{3.106}$$

the estimation

$$\|\Psi(k,k_o)\| \leq c \, \alpha^{k-k_o} \tag{3.107}$$

for $k \geq k_o$, where $\alpha := \beta + \delta c$, is valid.

PROOF: Let $\Phi(k,k_o)$ be the state transition matrix of the system $x(k+1)=A(k)x(k)$. If the term $A'(k)x(k)$ is interpreted as the input $f(k)$ of the system, then this solution for system (3.106) is obtained:

$$x(k) = \Phi(k,k_o)x_o + \sum_{i=k_o}^{k-1} \Phi(k,i+1)A'(i)x(i)$$

with the estimation

$$\|x(k)\| \leq c \, \beta^{k-k_o}\|x_o\| + \sum_{i=k_o}^{k-1} c \, \beta^{k-1-i} \|A'(i)\| \, \|x(i)\|.$$

This estimation multiplied by β^{-k} yields

$$\beta^{-k}\|x(k)\| \leq c \, \beta^{-k_o}\|x_o\| + \sum_{i=k_o}^{k-1} \frac{c}{\beta} \|A'(i)\| \, \beta^{-i} \|x(i)\|$$

or with

$$z(k) := \beta^{-k} \|x(k)\| , \tag{3.108}$$

$$z(k) \leq c\, z(k_o) + \sum_{i=k_o}^{k-1} \frac{c}{\beta} \|A'(i)\|\, z(i) .$$

From this one obtains with the help of Lemma 2-3

$$z(k) \leq c\, z(k_o) \prod_{i=k_o}^{k-1} [1 + \frac{c}{\beta}\|A'(i)\|]$$

$$\leq c\, z(k_o) \prod_{i=k_o}^{k-1} (1 + c\delta/\beta) ,$$

respective with (3.108)

$$\|x(k)\| \leq c\, \beta^{k-k_o}\|x_o\| \prod_{i=k_o}^{k-1} (1 + c\delta/\beta)$$

$$= c\, \beta^{k-k_o}\|x_o\|(1 + c\delta/\beta)^{k-k_o}$$

$$= c\, (\beta + c\delta)^{k-k_o}\|x_o\| . \tag{3.109}$$

With $\alpha := \beta + c\delta$ and $x(k) = \Psi(k,k_o)x_o$ from (3.109) the proposition (3.107) is following. $\blacksquare$

Theorem 3-60 and Definition 3-51 immediately yield the following theorem:

3-61   THEOREM: If (i) $A(k) \in E\mathcal{S}$, where $\|\Phi(k,k_o)\| \leq c\beta^{k-k_o}$,

                $c>0$, $\beta \in [0,1)$, and $k \geq k_o$,

        (ii) $\|A'(k)\| \leq \delta$, and

        (iii) $(\beta + c\delta) \in [0,1)$,

      then $[A(k) + A'(k)] \in E\mathcal{S}$.

3.7 Relations between the Stability Classes

In the preceding sections of this chapter for linear *time-variant* discrete-time systems

$$x(k+1) = A(k)\, x(k)$$

the following relations between the various stability classes were derived for $K=(\alpha,\infty)$:

(3.27):	$\mathbb{US} \subset \mathbb{S}$
Example 3.6:	$\mathbb{S} \smallsetminus \mathbb{US} \neq \emptyset$
"	$\mathbb{S} \smallsetminus \mathbb{AS} \neq \emptyset$
"	$\mathbb{AS} \neq \mathbb{US}$
(3.79):	$\mathbb{AS} \subset \mathbb{S}$
Lemma 3-47:	$\ell^P\mathbb{S} \subset \mathbb{S}$
Lemma 3-48:	$\ell^P\mathbb{S} \subset \mathbb{AS}$
Example 3.14:	$\mathbb{AS} \smallsetminus \ell^P\mathbb{S} \neq \emptyset$
Example 3.15:	$(\mathbb{US} \cap \mathbb{AS}) \smallsetminus \ell^P\mathbb{S} \neq \emptyset$
(3.90):	$\ell^P\mathbb{S} \cap \mathbb{US} \neq \emptyset$
Lemma 3-53:	$\mathbb{ES} = \ell^P\mathbb{S} \cap \mathbb{US}$
Lemma 3-58:	$\mathbb{ES} = \mathbb{UAS}$
(3.105):	$(\mathbb{US} \cap \mathbb{AS}) \smallsetminus \mathbb{ES} \neq \emptyset$

These relations are summarized and represented in Fig.3/13.

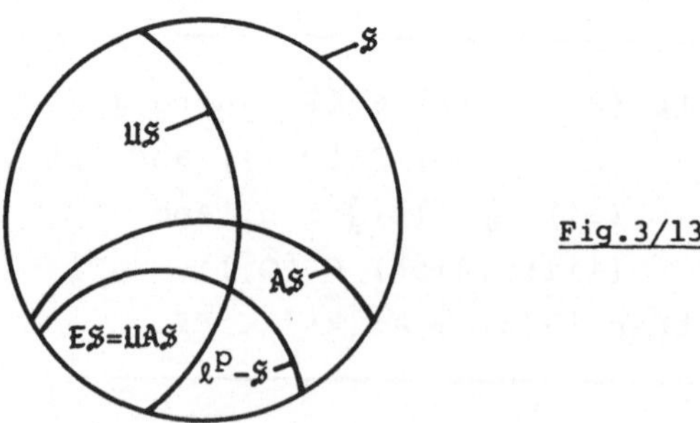

Fig.3/13

For linear *time-invariant* discrete-time systems

$$x(k+1) = A\ x(k)$$

and $K=(\alpha,\infty)$ these three relations were derived:

Section 3.3: $\mathfrak{U}\mathcal{S} = \mathcal{S}$

Lemma 3-49: $\ell^P\mathcal{S} \subset \mathfrak{U}\mathcal{S}$

Lemma 3-59: $A\mathcal{S} = E\mathcal{S} = \mathfrak{U}A\mathcal{S} = \ell^P\mathcal{S}$

They are visualized by <u>Fig.3/14</u>.

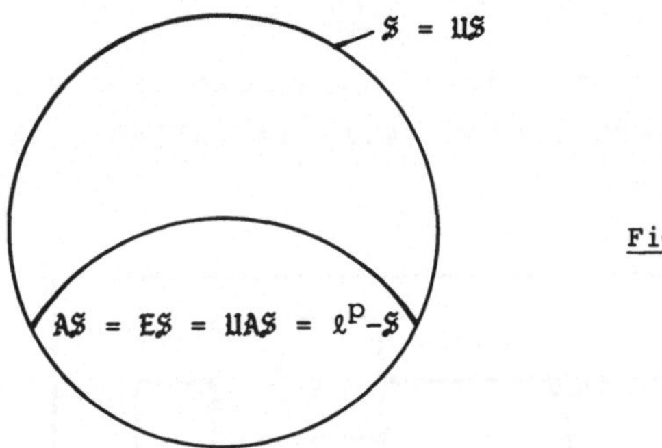

<u>Fig.3/14</u>

4 Stability of Forced Discrete-Time Systems

4.1 PRELIMINARY RESULTS

Chapter 3 dealt with dynamic discrete-time systems called *free* or *unforced*, because there was no input. A system with input is called a *forced* system. The mathematical description of a discrete-time system with input (see **Fig.4/1**) is

$$x(k+1) = A(k)\, x(k) + f(k). \tag{4.1}$$

A forced linear time-variant discrete-time system with the mathematical description (4.1) is called $\{A(k),f(k)\}$ in the following.

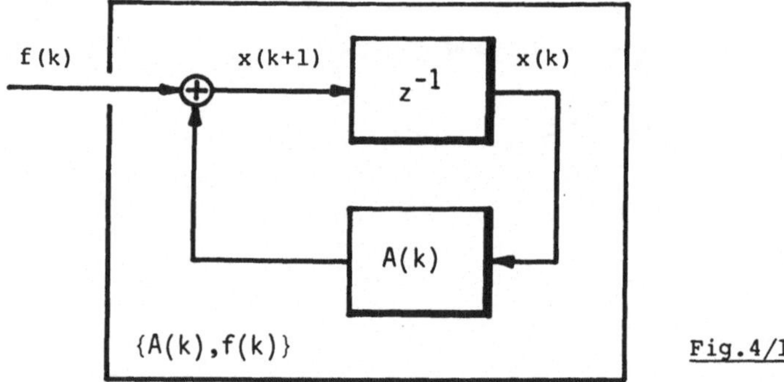

Fig.4/1

The solution of the state equation (4.1) for the initial state $x(k_o)=x_o$ is by Lemma 2-1

$$x(k) = \Phi(k,k_o)\, x_o + \sum_{i=k_o}^{k-1} \Phi(k,i+1)\, f(i). \tag{4.2}$$

If the system matrix A is time-invariant the solution for the initial time $k_o=0$ is in accordance to (2.7)

$$x(k) = A^k x_o + \sum_{i=0}^{k-1} A^{k-1-i} f(i). \tag{4.3}$$

For forced systems $\{A(k),f(k)\}$ the same different kinds
of stability definitions are used as for the free system
$x(k+1)=A(k)x(k)$ in the preceding chapter:

4-1   **DEFINITION:** The forced system $\{A(k),f(k)\}$ is said to be

      (i)   _stable_, if $A(k)\in S$,

      (ii)   _uniformly stable_, if $A(k)\in US$,

      (iii) _asymptotically stable_, if $A(k)\in AS$,

      (iv) _exponentially stable_, if $A(k)\in ES$,

      (v)   _bounded_ or _LAGRANGE-stable_, if

$$\forall_{k_o}\in K \; \exists \mu(k_o)>o \; \forall k\in[k_o,\omega) \; \forall f(.)$$

$$\|x(k;k_o,x_o,f(.))\| < \mu(k_o).$$

Some properties of the solution (4.2) depend on the
input sequence $f(.)$, while other properties are independent
of $f(.)$. If x' and x'' are two solutions of the state equation
(4.1), the solution equation (4.2) yields

$$x'(k) - x''(k) = \Phi(k,k_o) \, (x_o' - x_o''). \tag{4.4}$$

4-2   **THEOREM:** The solution x (and therefore every other
solution) of the forced system $\{A(k),f(k)\}$
has exactly the same stability properties as
the solutions of the free system $x(k+1)=A(k)x(k)$
except in the case of LAGRANGE-stability.

Definition 4-1 (v) means that $\{A(k),f(k)\}$ is said to be
bounded or LAGRANGE-stable, if all solutions are bounded.
From the solution (4.2) of $\{A(k),f(k)\}$ follows immediately:

$\underline{4-3}$

> **LEMMA:** The forced system $\{A(k),f(k)\}$ is bounded if and only if
> (i) $A(k) \in \mathcal{S}$, and
> (ii) $\{A(k),f(k)\}$ has a bounded solution. $\qquad$ (4.5)

In Lemma 4-3 condition (i) $A(k) \in \mathcal{S}$ alone don't suffice, since for example the forced system $x(k+1)=x(k)+1$ is stable but not bounded, because the solution $x(k)=x_0+k$ ($k_0=0$) is unbounded for $k \to \infty$. So the property *stability* is independent of the input sequence $f(.)$, but the property *boundedness* depends on $f(.)$.

Some definitions are needed for the following sections:

$\underline{4-4}$

> **DEFINITION:** (i) $\quad f(.) := \{f(i):i \in K\}.$ $\qquad$ (4.6)
>
> (ii) $\left\| f(.) \right\|_{p,K} := \left[\sum_{i \in K} \left\| f(i) \right\|^p \right]^{1/p}$, $p \in [1,\infty)$. $\qquad$ (4.7)
>
> (iii) $\left\| f(.) \right\|_{\infty,K} := \sup_{i \in K} \left\| f(i) \right\|.$ $\qquad$ (4.8)
>
> (iv) $\ell_K^p := \{f(.): \left\| f(.) \right\|_{p,K} < c < \infty\}$, $p \in [1,\infty].$ $\qquad$ (4.9)

If the class of systems in Lemma 4-3 is restricted, one has

$\underline{4-5}$

> **LEMMA:** If $A(k) \in \ell^p \mathcal{S}$ for a given $p \in [1,\infty]$, then the forced system $\{A(k),f(k)\}$ is bounded for every input sequence $f(.) \in \ell_K^q$, where $p^{-1}+q^{-1}=1$.

PROOF: For $k \geq k_0$ we have for the solution (4.2) with the HÖLDER inequality the estimate

$$\left\| x(k) \right\| \leq \left\| \Phi(k,k_0) \right\| \cdot \left\| x_0 \right\| + \sum_{i=k_0}^{k-1} \left\| \Phi(k,i+1) \right\| \cdot \left\| f(i) \right\| \leq$$

$$\leq \|\Phi(k,k_o)\| \cdot \|x_o\| + \left[\sum_{i=k_o}^{k-1} \|\Phi(k,i+1)\|^p\right]^{1/p} \left[\sum_{i=k_o}^{k-1} \|f(i)\|^q\right]^{1/q} . \quad (4.10)$$

Since $\ell^p \subsetneq \mathscr{S}$, Lemma 4-5 is following. $\blacksquare$

The opposite direction of the statement in Lemma 4-5 above shall be stated in Lemma 4-7 below. First the following Lemma 4-6 is needed to prove Lemma 4-7.

4-6

> LEMMA: If the forced system $\{A(k),f(k)\}$ is bounded for all $f(.) \in \ell_K^q$, then there exists a constant $M=M(k_o,q)$ such that for $k \in [k_o,\omega)$
>
> $$\left\|\sum_{i=k_o}^{k-1} \Phi(k,i+1)f(i)\right\| \leq M \|f(.)\|_{q,K} \qquad (4.11)$$
>
> is valid.

PROOF: With $\|f(.)\|_{q,K}$ according to definition (4.7) the estimation (4.10) yields

$$\|x(k)\| \leq \|\Phi(k,k_o)\| \cdot \|x_o\| + \left[\sum_{i=k_o}^{k-1} \|\Phi(k,i+1)\|^p\right]^{1/p} \|f(.)\|_{q,K}. \quad (4.12)$$

But $\|x(k)\|$ is bounded for all $k \in [k_o,\omega)$ only if

$$\left[\sum_{i=k_o}^{k-1} \|\Phi(k,i+1)\|^p\right]^{1/p}$$

is bounded for all $k \in [k_o,\omega)$, i.e., if a positive $M < \infty$ exists, such that

$$\left[\sum_{i=k_o}^{k-1} \|\Phi(k,i+1)\|^p\right]^{1/p} \leq M$$

is valid, thus (4.11) is valid. ∎

Now it can be formulated:

$\underline{4-7}$ | **LEMMA:** If the forced system $\{A(k), f(k)\}$ is bounded for every input sequence $f(.) \in \ell_K^q$ for a given $q \in [1, \infty]$, then $A(k) \in \ell^p \mathcal{S}$, where $p^{-1} + q^{-1} = 1$.

PROOF: The somewhat long proof is given indirect, i.e., not 'if A then B', but 'if not B then not A' is proved. So instead of Lemma 4-7 it shall be proved: if $A(k) \notin \ell^p \mathcal{S}$, then there exists an input sequence $f(.) \in \ell_K^q$ such that the forced system $\{A(k), f(k)\}$ is not bounded. If $A(k) \in \ell^p \mathcal{S}$, then by (3.60) there is a $k_o \in K$, such that for every $c > 0$ exists a $k_c > k_o$ and

$$\left[\sum_{i=k_o}^{k_c-1} \|\Phi(k_c, i+1)\|^p \right]^{1/p} > c \tag{4.13}$$

is valid. Now a sequence $f_c(.)$ is defined with the following properties:

(i) $f_c(.) \in \ell_K^q$,

(ii) $q^{-1} + p^{-1} = 1$, i.e., $q = p/(p-1)$,

(iii) $\|f(.)\|_{q,K} = 1$, and therefore

(iv) $\left\| \sum_{i=k_o}^{k_c-1} \Phi(k_c, i+1) f_c(i) \right\| \leq \left[\sum_{i=k_o}^{k_c-1} \|\Phi(k_c, i+1)\|^p \right]^{1/p}. \tag{4.14}$

Such a sequence $f_c(.)$ is e.g. ($x_o \in \mathbb{R}^n$, $x_o \neq 0$):

$$f_c(i) = \begin{cases} 0 & \text{for } i \in (k_c, \omega) \\[2ex] \left[\displaystyle\sum_{j=k_o}^{k_c-1} \|\Phi^*(k_c, j+1) x_o\|_2^p \right]^{-1/q} \|\Phi^*(k_c, i+1) x_o\|_2^{p-2} \Phi^*(k_c, i+1) x_o & \\[2ex] & \text{for } k_o \leq i \leq k_c. \end{cases}$$

Indeed it is

$$\|f(.)\|_{q,K} = \left\{\left[\sum_{j=k_o}^{k_c-1} \|\Phi^*(k_c,j+1)x_o\|_2^p\right]^{-1} \sum_{i=k_o}^{k_c-1} \left\|\|\Phi^*(k_c,i+1)x_o\|_2^{p-2} \times \right.\right.$$
$$\left.\left. \times \Phi^*(k_c,i+1)x_o\right\|_2^q\right\}^{1/q}$$
$$= \left\{\left[\sum_{j=k_o}^{k_c-1} \|\Phi^*(k_c,j+1)x_o\|_2^p\right]^{-1} \sum_{i=k_o}^{k_c-1} \|\Phi^*(k_c,i+1)x_o\|_2^{(p-1)q}\right\}^{1/q}$$
$$= \left\{\left[\sum_{j=k_o}^{k_c-1} \|\Phi^*(k_c,j+1)x_o\|_2^p\right]^{-1} \sum_{i=k_o}^{k_c-1} \|\Phi^*(k_c,i+1)x_o\|_2^p\right\}^{1/q}$$
$$= 1^{1/q} = 1. \tag{4.15}$$

With this sequence $f_c(.)$ it is

$$\sum_{i=k_o}^{k_c-1} \Phi(k_c,i+1)f_c(i) = \left[\sum_{i=k_o}^{k_c-1} \|\Phi^*(k_c,i+1)x_o\|_2^p\right]^{-1/q} \sum_{i=k_o}^{k_c-1} \Phi(k_c,i+1) \times$$
$$\times \|\Phi^*(k_c,i+1)x_o\|_2^{p-2} \Phi^*(k_c,i+1)x_o, \tag{4.16}$$

i.e., with $-q^{-1}=p^{-1}-1$

$$\sum_{i=k_o}^{k_c-1} \Phi(k_c,i+1)f_c(i) = \left[\sum_{i=k_o}^{k_c-1} \|\Phi^*(k_c,i+1)x_o\|_2^p\right]^{-1/p} \times$$
$$\times \left[\sum_{i=k_o}^{k_c-1} \|\Phi^*(k_c,i+1)x_o\|_2^p\right]^{-1} \sum_{i=k_o}^{k_c-1} \Phi(k_c,i+1)\|\Phi^*(k_c,i+1)x_o\|_2^{p-1} \times$$
$$\times \Phi^*(k_c,i+1)x_o. \tag{4.17}$$

Equation (4.17) multiplied from the left by x_o^* yields

$$\left| x_o^* \sum_{i=k_o}^{k_c-1} \Phi(k_c,i+1) f_c(i) \right| = \left[\sum_{i=k_o}^{k_c-1} \| \Phi^*(k_c,i+1) x_o \|_2^p \right]^{1/p} \times$$

$$\times \left[\sum_{i=k_o}^{k_c-1} \| \Phi^*(k_c,i+1) \|_2^p \right]^{-1} \sum_{i=k_o}^{k_c-1} \| \Phi^*(k_c,i+1) x_o \|^p$$

$$= \left[\sum_{i=k_o}^{k_c-1} \| \Phi^*(k_c,i+1) x_o \|_2^p \right]^{1/p}. \tag{4.18}$$

From (4.18) one has the estimation

$$\left[\sum_{i=k_o}^{k_c-1} \| \Phi^*(k_c,i+1) x_o \|_2^p \right]^{1/p} \leq \| x_o \|_2 \left\| \sum_{i=k_o}^{k_c-1} \Phi(k_c,i+1) f_c(i) \right\|_2, \tag{4.19}$$

hence for the norm of the state transition matrix

$$\left[\sum_{i=k_o}^{k_c-1} \| \Phi(k_c,i+1) \|^p \right]^{1/p} \leq \left\| \sum_{i=k_o}^{k_c-1} \Phi(k_c,i+1) f_c(i) \right\|_2. \tag{4.20}$$

On the other side it follows with the HÖLDER inequality and $\| f_c(\cdot) \|_{q,K} = 1$:

$$\left\| \sum_{i=k_o}^{k_c-1} \Phi(k_c,i+1) f_c(i) \right\| \leq \sum_{i=k_o}^{k_c-1} \| \Phi(k_c,i+1) f_c(i) \|$$

$$\leq \left[\sum_{i=k_o}^{\omega-1} \| \Phi(k_c,i+1) \|^p \right]^{1/p} \left[\sum_{i=k_o}^{\omega-1} \| f_c(i) \|^q \right]^{1/q}$$

$$= \left[\sum_{i=k_o}^{\omega-1} \| \Phi(k_c,i+1) \|^p \right]^{1/p}. \tag{4.21}$$

But (4.20) and (4.21) yields (4.14). Putting (4.14)
into (4.13) it follows that for the choosen bounded
sequence $f_c(.)$ the forced system $\{A(k),f(k)\}$ is not
bounded. ∎

Now Lemma 4-5 combined with Lemma 4-7 yields:

4-8 THEOREM: For a given $p\in[1,\infty]$ is $A(k)\in\ell^p\mathcal{S}$ if and only
if for every sequence $f(.)\in\ell_K^q$, where $q^{-1}+p^{-1}=1$,
the forced system $\{A(k),f(k)\}$ is bounded.

With $\mathfrak{ES}=\ell^p\mathcal{S}\cap\ell\mathcal{S}$ from Lemma 3-53 it follows from Theorem 4-8:

4-9 THEOREM: $A(k) \in \mathfrak{ES}$ if and only if the forced system
$\{A(k),f(k)\}$ is bounded for every sequence
$f(.) \in \ell_K^q$ and every $q\in[1,\infty]$.

4.2 INPUT-STATE STABILITY

Scientific and technical systems usually have less input
variables than state variables, viz. they have mathematical
descriptions of the following form

$$x(k+1) = A(k)\ x(k) + B(k)\ u(k)\ , \qquad (4.22)$$

where $x(k)\in\mathbb{C}^n$, $u(k)\in\mathbb{R}^m$, $A(k)\in\mathbb{C}^{n\times n}$, $B(k)\in\mathbb{C}^{n\times m}$, and $m\leq n$. A
forced linear time-variant discrete-time system with the
mathematical description (4.22) is designated $\{A(k),B(k)\}$.

The different kinds of vector norms for the input vector
have in technical systems a quite natural meaning:

$$\|u(k)\|_1 = \sum_{i=1}^{m} |u_i(k)| \quad \text{is proportional to the} \qquad (4.23)$$
$$\text{fuel input,}$$

$$\|u(k)\|_2 = \left[\sum_{i=1}^{m} |u_i(k)|^2\right]^{1/2} \quad \text{is proportional to the} \quad (4.24)$$

input energy,

$$\|u(k)\|_\infty = \sup_{i} |u_i(k)| \qquad \text{is the least upper bound} \quad (4.25)$$

for the maximal input amplitude $u_i(k)_{max}$.

If in $\|u(.)\|_{p,K}$ not the $\|.\|_2$-norm but the associated vector norm $\|.\|_p$ is used, i.e., it is for $p \in [1,\infty]$ defined

$$\|u(.)\|_{p,K}^* := \left[\sum_{k \in K} \|u(k)\|_p^p\right]^{1/p} =$$

$$= \left[\sum_{k \in K} \sum_{i=1}^{m} |u_i(k)|^p\right]^{1/p} \qquad (4.26)$$

then it is

$$\|u(.)\|_{1,K}^* = \sum_{k \in K} \sum_{i=1}^{m} |u_i(k)| \quad \text{is proportional to the} \quad (4.27)$$

total fuel consumption over K,

$$\|u(.)\|_{2,K}^* = \left[\sum_{k \in K} \sum_{i=1}^{m} |u_i(k)|^2\right]^{1/2} \quad \text{is proportional} \quad (4.28)$$

to the total amount of energy over K, and

$$\|u(.)\|_{\infty,K}^* = \sup_{j \in K} \sup_{i \in [1,m]} |u_i(j)| \quad \text{is proportional to} \quad (4.29)$$

the least upper bound for the maximal input amplitude $|u_i(k)|_{max}$ over K.

The last remarks above are motivating following definition:

4-10   DEFINITION: The system $\{A(k),B(k)\}$ is said to be ℓ^p-*BIBS-stable*, $p\in[1,\infty]$, if $u(.)\in\ell_K^p$ implies $x(.)\in\ell_K^p$.

BIBS is an abbreviation for *Bounded-Input/Bounded-State*. As usual for $p=\infty$ especially it is defined:

4-11   DEFINITION: The system $\{A(k),B(k)\}$ is said to be *BIBS-stable*, if it is ℓ^∞-BIBS-stable.

BIBS-stability may be defined with other words in detail in this way:

4-12   DEFINITION: The system $\{A(k),B(k)\}$ is said to be *BIBS-stable*, if for all bounded input sequences $u(.)$ ($\|u(k)\|<M_1<\infty$ for all $k\in K$) the state vector sequence $x(.)$ is bounded too ($\|x(k)\|<M_2<\infty$ for all $k\in K$).

For the solution (see Lemma 2-1)

$$x(k) = \Phi(k,k_0)x_0 + \sum_{i=k_0}^{k-1} \Phi(k,i+1)B(i)u(i) \qquad (4.30)$$

of the state equation (4.22) by the HÖLDER inequality the estimate

$$\|x(k)\| \leq \|\Phi(k,k_0)x_0\| + \left[\sum_{i=k_0}^{k-1} \|\Phi(k,i+1)B(i)\|^p\right]^{1/p}\left[\sum_{i=k_0}^{k-1} \|u(i)\|^q\right]^{1/q}$$

$$(4.31)$$

is obtained. If the input sequence u(.) is bounded, i.e., if

$$\left[\sum_{i \in K} \| u(i) \|^{\infty} \right]^{1/\infty} = \sup_{i} \| u(i) \|$$

is finite, and the initial state x_o is equal to the zero state, one receives from (4.31) immediately for p=1:

4-13   **LEMMA:** The system $\{A(k),B(k)\}$ is BIBS-stable for $x_o=0$, if there exists a finite constant $c<\infty$, such that for all $k_o \in K, k \in [k_o,\omega)$

$$\sum_{i=k_o}^{k-1} \| \Phi(k,i+1)B(i) \| < c. \tag{4.32}$$

If the matrix sequence B(.) is bounded, i.e., if there exists an upper bound $c<\infty$ such that $\sup_{i \in K}\|B(i)\|=c$, then it follows with Theorem 3-60:

4-14   **LEMMA:** If $A(k) \in \mathbf{E\mathscr{S}}$ and B(.) is bounded, $B(.) \in \ell_K^{\infty}$, then the system $\{A(k),B(k)\}$ is BIBS-stable.

This lemma also would be received from Theorem 4-9. But from Theorem 4-9 immediately the main theorem is following:

4-15   **THEOREM:** If B(k) is bounded and $A(k) \in \ell^p\mathscr{S}$, then the system $\{A(k),B(k)\}$ is ℓ^q-BIBS-stable, where $q^{-1} + p^{-1} = 1$.

Since $\mathbf{E\mathscr{S}} \subset \ell^p\mathscr{S}$ in particular one obtains:

$$
\underline{4\text{-}16} \quad \boxed{\begin{array}{l} \text{THEOREM: If B(k) is bounded and A(k)} \in \boldsymbol{ES}\text{, then the} \\ \qquad\qquad \text{system } \{A(k),B(k)\} \text{ is } \ell^P\text{-BIBS-stable for} \\ \qquad\qquad \text{every } p\in[1,\infty]. \end{array}}
$$

The converse direction of Theorem 4-16 is not always true. ℓ^P-BIBS-stability does not imply always exponential stability of the free system. The relation between internal and external stability would be the main topic of chapter 5.

4.3 INPUT-OUTPUT STABILITY

In this section the influence of the input sequence u(.) on the output sequence y(.) of the linear time-variant discrete-time system with the mathematical description

$$x(k+1) = A(k)\ x(k) + B(k)\ u(k)\ , \tag{4.33}$$

$$y(k) = C(k)\ x(k)\ , \tag{4.34}$$

is considered (see <u>Fig.4/2</u>). A system with above mentioned mathematical description is designated $\{A(k),B(k),C(k)\}$.

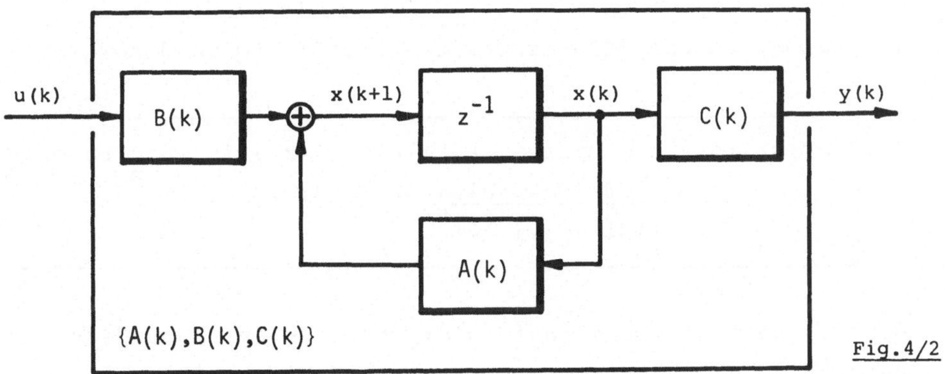

<u>Fig.4/2</u>

The solution of the state equation (4.33) for the initial state $x(k_o)=x_o$ is in according to Lemma 2-1

$$x(k) = \Phi(k,k_o)x_o + \sum_{i=k_o}^{k-1} \Phi(k,i+1)B(i)u(i) \tag{4.35}$$

so that the output is given by (4.34) as

$$y(k) = C(k)\Phi(k,k_o)x_o + \sum_{i=k_o}^{k-1} C(k)\Phi(k,i+1)B(i)u(i). \tag{4.36}$$

If the system $\{A(k),B(k),C(k)\}$ is initially at k_o relaxed $(x_o=O)$, with the *impulse response matrix*

$$G(k,i) := C(k)\Phi(k,i+1)B(i) \tag{4.37}$$

one has the input-output description

$$y(k) = \sum_{i=k_o}^{k-1} G(k,i)u(i) , \tag{4.38}$$

where $y(k)\in R^r$, $u(i)\in R^m$, $G(k,i)\in R^{r\times m}$, and

$$G(k,i) = O \quad \text{for } i>k , \tag{4.39}$$

i.e., the system is supposed to be not anticipative.

In analogy to the BIBS-stability it is introduced now

4-17   **DEFINITION**: the system $\{A(k),B(k),C(k)\}$ is said to be ℓ^p-*BIBO-stable*, $p\in[1,\infty]$, if $u(.)\in\ell_K^p$ implies $y(.)\in\ell_K^p$.

BIBO is here an abbreviation for *Bounded-Input/Bounded-Output*. As usual the special type of ℓ^∞-BIBO-stability is defined in this way:

$$\underline{4\text{-}18} \quad \boxed{\begin{array}{l} \text{DEFINITION: The system } \{A(k),B(k),C(k)\} \text{ is said to be} \\ \qquad\qquad BIBO\text{-}stable, \text{ if it is } \ell^{\infty}\text{-BIBO-stable.} \end{array}}$$

With other words, BIBO-stability means, that any bounded input sequence produces a bounded output sequence, or if $\|u(k)\| < M_1 < \infty$, then $\|y(k)\| < M_2 < \infty$ for all $k \in K$.

The input-output description (4.38) is yielding the estimation

$$\|y(k)\| \leq \sum_{i=k_0}^{k-1} \|G(k,i)\|\,\|u(i)\|$$

$$\leq c_1 \sum_{i=k_0}^{k-1} \|G(k,i)\| . \tag{4.40}$$

Thus it is:

$$\underline{4\text{-}19} \quad \boxed{\begin{array}{l} \text{LEMMA: The relaxed system } \{A(k),B(k),C(k)\} \text{ is BIBO-} \\ \qquad\text{stable, if there exists a constant } c<\infty \text{ such} \\ \qquad\text{that for all } k_0 \in K \text{ and } k \in [k_0,\omega) \\[2mm] \qquad \displaystyle\sum_{i=k_0}^{k-1} \|G(k,i)\| < c \qquad\qquad (4.41) \\[2mm] \qquad\text{is valid.} \end{array}}$$

Condition (4.41) is not only sufficient, but also necessary:

$$\underline{4\text{-}2o} \quad \boxed{\begin{array}{l} \text{THEOREM: The relaxed system } \{A(k),B(k),C(k)\} \text{ is BIBO-} \\ \qquad\qquad\text{stable, if and only if there exists a constant} \\ \qquad\qquad c<\infty \text{ such that} \end{array}}$$

$$\sum_{i=\alpha+1}^{\omega-1} \|G(k,i)\| < c \qquad (4.42)$$

is valid for all $k \in K$.

PROOF: 1.($\Leftarrow$)I.e., first is proved: (4.42) implies BIBO-stability. Given $u(.) \in \ell_K^\infty$, the equations (4.38) and (4.39) yield

$$\|y(k)\| \leq \sum_{i=k_o}^{k-1} \|G(k,i)\| \, \|u(i)\|$$

$$\leq \sum_{i=\alpha+1}^{\omega-1} \|G(k,i)\| \, \|u(i)\|$$

$$\leq c \cdot c_1$$

$$\leq c_s \cdot c_1 \qquad (4.43)$$

where $c_s := \sup_{k \in K} c$, such that the right hand side of

(4.43) is independent of k.

2.($\Rightarrow$)I.e., BIBO-stability implies (4.42). Contradiction assumption: the system is BIBO-stable, but for someone k_o there exists for arbitrary c' a $k' \in [k_o,\omega)$, such that

$$\sum_{i=k_o}^{k'-1} \|G(k',i)\| > c', \qquad (4.44)$$

that is, (4.42) is not valid. For vectors v, w, and y and a matrix M the following is valid

$$\|y\| = \max_{\|v\|=1} v^T y \qquad (4.45)$$

and

$$\|M\| = \max_{\|w\|=1} \|Mw\| = \max_{\|v\|=1} \max_{\|w\|=1} v^T Mw. \qquad (4.46)$$

From (4.45) with (4.38) for $k=k'$ follows

$$\|y(k')\| = \max_{\|v\|=1} v^T \left[\sum_{i=k_o}^{k'-1} G(k',i)u(i) \right]$$

$$= \sum_{i=k_o}^{k'-1} \max_{\|v\|=1} v^T G(k',i)u(i). \qquad (4.47)$$

If for bounded $\|u(i)\| \leq 1$ that input sequence $u(.)$ is choosen for which the term in the sum (4.47) is maximum, equation (4.47) together with (4.46) yields

$$\|y(k')\| = \sum_{i=k_o}^{k'-1} \max_{\|v\|=1} \max_{\|u(i)\|=1} v^T G(k',i)u(i)$$

$$= \sum_{i=k_o}^{k'-1} \|G(k',i)\|. \qquad (4.48)$$

Hence $\|y(k')\|$ cannot be bounded by the contradiction assumption although the input sequence $u(.)$ was bounded by $\|u(i)\| \leq 1$. This is completing the proof of Theorem 4-20. ∎

For nonanticipative systems $G(k,i)=0$ is valid for $i>k$. So one obtains with $C(k) \equiv I$ from Theorem 4-20 a theorem, which is an extension of Lemma 4-13:

4-21 THEOREM: The relaxed system $\{A(k),B(k)\}$ is BIBS-stable, if and only if there exists a positive constant $c<\infty$ such that

$$\sum_{i=\alpha+1}^{k-1} \|\Phi(k,i+1)B(i)\| < c \qquad (4.49)$$

is valid for all $k \in K$.

A connection between exponential stability and ℓ^p-BIBO-stability is stated in the following theorem.

<u>4-22</u> THEOREM: If the system $\{A(k),B(k),C(k)\}$ has bounded matrices $A(k)$, $B(k)$, and $C(k)$ and $A(k) \in E\mathcal{S}$, then the system is ℓ^p-BIBO-stable.

PROOF: With regard to the boundedness assumption in the theorem, there is a constant $c < \infty$ such that

$$\|A(k)\| < c, \quad \|B(k)\| < c, \quad \text{and} \quad \|C(k)\| < c \qquad (4.50)$$

for all $k \in K$ and since $A(k) \in E\mathcal{S}$ by Theorem 3-54 is

$$\sum_{i=k_0}^{k-1} \|\Phi(k,i+1)\| < c \qquad (4.51)$$

for all $k_0 \in K$ and $k \in [k_0, \omega)$. For $y(k)$ one has the estimation

$$\|y(k)\| \leq \sum_{i=\alpha+1}^{k-1} \|G(k,i)\| \, \|u(i)\|$$

$$\leq \sum_{i=\alpha+1}^{k-1} \|G(k,i)\|^{1/p} \|u(i)\| \, \|G(k,i)\|^{1/q} \qquad (q^{-1}+p^{-1}=1)$$

$$\leq \left[\sum_{i=\alpha+1}^{k-1} \|G(k,i)\| \, \|u(i)\|^{p} \right]^{1/p} \left[\sum_{i=\alpha+1}^{k-1} \|G(k,i)\| \right]^{1/q} \qquad (\text{HÖLDER-ine.})$$

$$\leq \left[\sum_{i=\alpha+1}^{k-1} \|G(k,i)\| \|u(i)\|^p \right]^{1/p} \left[c^2 \sum_{i=\alpha+1}^{k-1} \|\Phi(k,i+1)\| \right]^{1/q} \qquad \text{(because of 4.50)}$$

$$\leq c^{3/q} \left[\sum_{i=\alpha+1}^{k-1} \|G(k,i)\| \|u(i)\|^p \right]^{1/p}. \qquad \text{(because of 4.51)}$$

So it is

$$\|y(.)\|_{p,K}^p \leq c^{3p/q} \sum_{k=\alpha+1}^{\omega-1} \left[\sum_{i=\alpha+1}^{k-1} \|G(k,i)\| \|u(i)\|^p \right]$$

$$\leq c^{3p/q} \sum_{i=\alpha+1}^{\omega-1} \|u(i)\|^p \sum_{k=\alpha+1}^{\omega-1} \|G(k,i)\|$$

$$\leq c^{3p/q} c^3 \sum_{i=\alpha+1}^{\omega-1} \|u(i)\|^p$$

$$\leq c^{3p} \|u(.)\|_{p,K}^p,$$

hence

$$\|u(.)\|_{p,K} < c_1 < \infty \quad \Rightarrow \quad \|y(.)\|_{p,K} < c_2 < \infty. \blacksquare$$

4.4 Short Time Boundedness of Forced Systems

In Section 3.2 short time boundedness for given bounds was introduced for *unforced* linear time-variant discrete-time systems. Now this kind of boundedness shall be stated for *forced* systems.

4-23   **DEFINITION:** The system $\{A(k), f(k)\}$ is said to be *short time BIBS-bounded*, if for *specified* finite $\delta_o > 0$, $\delta_1 > 0$, $\varepsilon > 0$, and $N > 0$ for every

$$k_o \in (\alpha, \omega - N), \quad \|x_o\| < \delta_o \quad \text{and} \quad \|f(k)\| < \delta_1 \quad \text{implies}$$

$$\|x(k;,k_o,x_o,f(.))\| < \varepsilon \qquad (4.52)$$

$$\text{for all } k \in [k_o, k_o + N].$$

Two sufficient conditions for short time BIBS-boundedness are given in the following two theorems.

4-24   THEOREM: The system $\{A(k), f(k)\}$ is short time BIBS-bounded for specified δ_o, δ_1, ε, and N, if for all $k \in [k_o, k_o + N]$

$$\delta_o \prod_{i=k_o}^{k-1} \lambda_{Hmax}^{1/2}(i) + \delta_1 \sum_{i=k_o}^{k-1} \prod_{j=i+1}^{k-1} \lambda_{Hmax}^{1/2}(i) < \varepsilon \qquad (4.53)$$

is valid, where λ_{Hmax} is the maximal eigenvalue of $H(k) = A^*(k)A(k)$.

PROOF: For the solution

$$x(k) = \Phi(k, k_o) x_o + \sum_{i=k_o}^{k-1} \Phi(k, i+1) f(i)$$

of the system $\{A(k), f(k)\}$ one gets the estimation

$$\|x(k)\| \leq \|\Phi(k, k_o)\| \, \|x_o\| + \sum_{i=k_o}^{k-1} \|\Phi(k, i+1)\| \, \|f(i)\|$$

and with the estimation (2.31) for the state transition matrix

$$\|x(k)\| \leq \prod_{i=k_o}^{k-1} \lambda_{Hmax}^{1/2}(i) \, \|x_o\| + \sum_{i=k_o}^{k-1} \prod_{j=i+1}^{k-1} \lambda_{Hmax}^{1/2}(j) \, \|f(i)\| . \qquad (4.54)$$

Since $\|x_o\| < \delta_o$ and $\|f(i)\| < \delta_1$ are presupposed from 4.54 follows

$$\|x(k)\| < \delta_o \prod_{i=k_o}^{k-1} \lambda_{Hmax}^{1/2}(i) + \delta_1 \sum_{i=k_o}^{k-1} \prod_{j=i+1}^{k-1} \lambda_{Hmax}^{1/2}(j).$$

If the right-hand side is less than ε, then $\|x(k)\| < \varepsilon$. ∎

Using the estimate (2.34a)

$$\|\Phi(k,k_o)\| \le \exp\left[\sum_{i=k_o}^{k-1} \ln\|A(i)\|\right]$$

for the state transition matrix, the following theorem can be proved in the same way as Theorem 4-24 above.

4-25 THEOREM: The system $\{A(k),f(k)\}$ is short time BIBS-bounded for specified δ_o, δ_1, ε, and N, if for all $k_o \in (\alpha, \omega-N)$ and all $k \in [k_o, k_o+N]$

$$\delta_o \exp\left[\sum_{i=k_o}^{k-1} \ln\|A(i)\|\right] + \delta_1 \sum_{i=k_o}^{k-1} \exp\left[\sum_{j=i+1}^{k-1} \ln\|A(j)\|\right] < \varepsilon \tag{4.55}$$

is valid.

For the system $\{A(k),B(k),C(k)\}$ with input and output variables, following kind of boundedness is introduced:

4-26 DEFINITION: The system $\{A(k),B(k),C(k)\}$ is said to be *short time BIBO-bounded*, if for specified $\delta>0$, $\varepsilon>0$, and $N>0$ for $x_o=0$ and every $k_o \in (\alpha, \omega-N)$, $\|u(k)\| < \delta$ implies $\|y(k)\| < \varepsilon$ for all $k \in [k_o, k_o+N]$.

The following theorem gives necessary and sufficient
conditions for this kind of boundedness.

$\underline{4\text{-}27}$ | THEOREM: The system $\{A(k),B(k),C(k)\}$ is short time
BIBO-bounded for specified δ, ε, and N, if
and only if for all $k_o \in (\alpha,\omega-N)$ and $k \in [k_o,k_o+N]$

$$\sum_{i=k_o}^{k-1} \|G(k,i)\| < \frac{\varepsilon}{\delta} \qquad (4.56)$$

is valid.

PROOF: 1.($\Leftarrow$) The zero state output of the system is given by

$$y(k) = \sum_{i=k_o}^{k-1} C(k)\Phi(k,i+1)B(i)u(i)$$

$$= \sum_{i=k_o}^{k-1} G(k,i)u(i). \qquad (4.57)$$

Therefore

$$\|y(k)\| \leq \sum_{i=k_o}^{k-1} \|G(k,i)\|\,\|u(i)\|$$

and since $\|u(i)\| < \delta$ is specified,

$$\|y(k)\| \leq \delta \sum_{i=k_o}^{k-1} \|G(k,i)\|.$$

Hence if (4.56) is valid, then $\|y(k)\| < \varepsilon$.

2.($\Rightarrow$) Contradiction assumption: the system is short
time BIBO-bounded, but for someone k_o there exists a
$k_1 \in [k_o,k_o+N]$, such that

$$\sum_{i=k_O}^{k_1-1} \| G(k_1,i) \| > \frac{\varepsilon}{\delta} \ . \tag{4.58}$$

From (4.45) follows with (4.57) for $k=k_1$

$$\| y(k_1) \| = \max_{\|v\|=1} v^T \left[\sum_{i=k_O}^{k_1-1} G(k_1,i)u(i) \right]$$

$$= \sum_{i=k_O}^{k_1-1} \max_{\|v\|=1} v^T G(k_1,i)u(i) \ . \tag{4.59}$$

If for bounded $\|u(k)\| \leq \delta$ that input sequence $u(.)$ is chosen, for which the term in the sum (4.59) is maximum, then equation (4.59) together with (4.46) yields

$$y(k_1) = \sum_{i=k_O}^{k_1-1} \max_{\|v\|=1} \max_{\|u(i)\|=\delta} v^T G(k_1,i)u(i)$$

$$= \delta \sum_{i=k_O}^{k_1-1} \| G(k_1,i) \| > \delta \frac{\varepsilon}{\delta} = \varepsilon \ .$$

Hence $\| y(k_1) \|$ cannot be bounded by the contradiction assumption, although the input sequence $u(.)$ was bounded by $\|u(k)\| \leq \delta$. ∎

5 Relations between Internal and External Stability

5.1 REACHABILITY

In control theory it is important to know whether a system
has this property: given the system in the zero state, can a
finite control sequence be found, which steers the system to
any given state at time k?

5-1 DEFINITION: A *state* x of a discrete-time system is said
to be *reachable at time k*, if there exists
a finite integer N>O and a finite input
sequence

$$u_{[k-N,k-1]} := \{u(k-N,u(k-N+1),...,u(k-1)\} \qquad (5.1)$$

which transfers the state x(k-N)=O to the
state

$$x(k) = x(k;k-N,O,u_{[k-N,k-1]}) = x. \qquad (5.2)$$

A *system* is said to be *reachable at k*, if
every state $x \in \mathbb{C}^n$ is reachable at k.
A *system* is said to be *reachable*, if the
system is reachable at every time $k \in K$.

A necessary and sufficient condition for the reachability
of a linear time-variant discrete-time system is

5-2 LEMMA: The system $\{A(k),B(k)\}$ of order n with the state
transition equation

$$x(k+1) = A(k)x(k) + B(k)u(k) \qquad (5.3)$$

---

is reachable at k, if and only if for a finite integer $N > 0$

$$\text{rank } S(k,N) = n, \tag{5.4}$$

where

$$S(k,N) := [B(k-1), \Phi(k,k-1)B(k-2), \ldots, \Phi(k,k-N+1)B(k-N)] \tag{5.5}$$

is the so called *reachability matrix*.

PROOF: 1.($\Leftarrow$) If $x(k-N) = 0$, then

$$x(k) = \sum_{i=k-N}^{k-1} \Phi(k,i+1)B(i)u(i)$$

$$= S(k,N)u(k,N), \tag{5.6}$$

where

$$u(k,N) := \begin{bmatrix} u(k-1) \\ u(k-2) \\ \vdots \\ u(k-N) \end{bmatrix}.$$

The input sequence

$$u(k,N) = S^T(k,N)[S(k,N)S^T(k,N)]^{-1}x \tag{5.7}$$

transfers the zero state to the state x at time k, while with (5.6)

$$x(k) = S(k,N)S^T(k,N)[S(k,N)S^T(k,N)]^{-1}x = 0;$$

the product matrix $S(k,N)S^T(k,N)$ is regular because rank $S(k,N) = n$.

2.($\Rightarrow$) Contradiction assumption: if the system is reachable at k, then

rank $S(k,N) < n$. $\hspace{6cm}$ (5.8)

If (5.8) is valid, then there exists a state $x(k) \neq 0$ such that

$$x^T(k)S(k,N) = 0^T.$$

This leads with (5.6) to

$$x^T(k)x(k) = x^T(k)S(k,N)u(k,N) = 0,$$

but this is a contradiction to the assumption $x(k) \neq 0$. Therefore (5.4) is a necessary condition for the reachability of the system. ∎

If the $n \times m \cdot N$ reachability matrix $S(k,N)$ has rank equal to n, then the $n \times n$ *reachability gramian*

$$W(k,k-N) := S(k,N)S^T(k,N)$$

$$= \sum_{i=k-N}^{k-1} \Phi(k,i+1)B(i)B^T(i)\Phi^T(k,i+1) \qquad (5.9)$$

has rank n too, is symmetric, and positive definite, such that from Lemma 5-2 follows:

<u>5-3</u> **LEMMA:** The system $\{A(k),B(k)\}$ of order n with the state transition equation

$$x(k+1) = A(k)x(k) + B(k)u(k) \qquad (5.10)$$

is reachable at k, if and only if for a finite integer $N>0$ the symmetric reachability gramian

$$W(k,k-N) := \sum_{i=k-N}^{k-1} \Phi(k,i+1)B(k)B^T(i)\Phi^T(k,i+1) \tag{5.11}$$

is positive definite:

$$W(k,k-N) > 0. \tag{5.12}$$

If $W(k,k-N)$ is positive definite, and if $A(i)$, $B(i)$ are bounded above, then the inverse matrix $W^{-1}(k,k-N)$ exists, and according to (5.7) an input sequence immediately can be constructed, such that *every* initial state $x_0 = x(k-N)$ is transferred to every final state $x_f = x(k)$. For $x_0 = x(k-N)$ the state transition equation (5.10) has the solution

$$x_f = x(k) = \Phi(k,k-N)x_0 + \sum_{i=k-N}^{k-1} \Phi(k,i+1)B(i)u(i)$$

$$= \Phi(k,k-N)x_0 + S(k,N)u(k,N), \tag{5.13}$$

where the input sequence

$$u(k,N) = S^T(k,N)W^{-1}(k,k-N)[x_f - \Phi(k,k-N)x_0], \tag{5.14}$$

i.e., for $j\in[k-N,k-1]$

$$u(j) = B^T(j)\Phi^T(k,j)W^{-1}(k,k-N)[x_f - \Phi(k,k-N)x_0], \tag{5.15}$$

steers the system to the state x_f at time k, because with (5.13)

$$x(k) = \Phi(k,k-N)x_0 + S(k,N)u(k,N)$$

$$= \Phi(k,k-N)x_0 + S(k,N)S^T(k,N)W^{-1}(k,k-N)[x_f - \Phi(k,k-N)x_0]$$

$$= \Phi(k,k-N)x_0 + W(k,k-N)W^{-1}(k,k-N)[x_f - \Phi(k,k-N)x_0]$$

$$= x_f. \tag{5.16}$$

In technical systems the quantity $u^T(k)u(k)$ is often proportional to the input energy at time k, such that

$$\|u(k,N)\|^2 = u^T(k,N)u(k,N)$$

$$= \sum_{i=k-N}^{k-1} u^T(i)u(i) \tag{5.17}$$

can be considered as the amount of energy which is spent in the transfer from x_o to x_f, by means of u(k,N). The input sequence given by (5.14), resp. (5.15), has an additional property:

5-4   LEMMA: Let the system $\{A(k),B(k)\}$ be reachable. Then for every $x_o,x_f \in \mathbb{C}^n$ the input sequence u(k,N) defined by (5.14), (5.15) resp., is among those input sequences $\bar{u}_{[k-N,k-1]}$, which transfers the system from the state x_o to the final state x_f, whose norm $\|u(k,N)\|^2$ is minimum.

PROOF: With $\bar{x}:=x_f-\Phi(k,k-N)x_o$, the input sequence (5.14) can be written

$$u(k,N) = S^T(k,N)W^{-1}(k,k-N)\bar{x}$$

$$= S^T(k,N)[S(k,N)S^T(k,N)]^{-1}\bar{x}. \tag{5.18}$$

Equation (5.18) multiplied from the left by S(k,N) gives

$$S(k,N)u(k,N) = \bar{x}. \tag{5.19}$$

If $\bar{u}(k,N)$ is a transfer input sequence too, it is again

$$S(k,N)\bar{u}(k,N) = \bar{x}. \tag{5.20}$$

Subtracting (5.20) from (5.19) leads to

$$S(k,N)[u(k,N)-\bar{u}(k,N)] = 0. \tag{5.21}$$

With (5.18) and (5.21) one gets for the scalar product

$$\{S(k,N)[u(k,N)-\bar{u}(k,N)]\}^T\{W^{-1}(k,k-N)\bar{x}\} = \tag{5.22}$$

$$= [u^T(k,N)-\bar{u}^T(k,N)]S^T(k,N)W^{-1}(k,k-N)\bar{x}$$

$$= [u^T(k,N)-\bar{u}^T(k,N)]u(k,N)$$

$$= 0, \tag{5.23}$$

that is

$$\|u(k,N)\|^2 = \bar{u}^T(k,N)u(k,N). \tag{5.24}$$

With

$$\|\bar{u}(k,N)\|^2 - \|\bar{u}(k,N)-u(k,N)\|^2 =$$

$$= \bar{u}^T(k,N)\bar{u}(k,N)-[\bar{u}(k,N)-u(k,N)]^T[\bar{u}(k,N)-u(k,N)]$$

$$= \bar{u}^T(k,N)\bar{u}(k,N)-\bar{u}^T(k,N)\bar{u}(k,N)+\bar{u}^T(k,N)u(k,N)+$$

$$+ u^T(k,N)\bar{u}(k,N)-u^T(k,N)u(k,N)$$

$$= \bar{u}^T(k,N)u(k,N) + u^T(k,N)[\bar{u}(k,N)-u(k,N)]$$

$$= \bar{u}^T(k,N)u(k,N) \tag{5.25}$$

equation (5.24) yields

$$\|u(k,N)\|^2 = \|\bar{u}(k,N)\|^2 - \|\bar{u}(k,N)-u(k,N)\|^2, \tag{5.26}$$

which proves, that $u(k,N)$ has minimum norm $\|u(k,N)\|$.
In fact, if

$$\|\bar{u}(k,N)\|^2 \leq \|u(k,N)\|^2,$$

from (5.26) follows

$$\|\bar{u}(k,N) - u(k,N)\|^2 = 0,$$

hence $\bar{u}(k,N)=u(k,N)$. $\blacksquare$

From (5.14) the minimal transfer energy follows additional:

$$\|u(k,N)\|^2 = \bar{x}^T W^{-1}(k,k-N) S(k,N) S^T(k,N) W^{-1}(k,k-N)\bar{x}$$

$$= \bar{x}^T W^{-1}(k,k-N)\bar{x}. \tag{5.27}$$

If

$$W(k,k-N) \geq \alpha I > 0, \tag{5.28}$$

then also the equivalent relation

$$W^{-1}(k,k-N) \leq \frac{1}{\alpha} I \tag{5.29}$$

is valid. Relation (5.29) is an abbreviation for the following associated relation for the quadratic forms

$$\bar{x}^T W^{-1}(k,k-N)\bar{x} \leq \frac{1}{\alpha} \bar{x}^T\bar{x}, \tag{5.30}$$

or with (5.27)

$$\|u(k,N)\|^2 \leq \frac{1}{\alpha}\|\bar{x}\|^2. \tag{5.31}$$

Inequality (5.28) in conjunction with inequality (5.31) is motivating the following definition:

<u>5-5</u> **DEFINITION:** The system $\{A(k),B(k)\}$ with the state equation

$$x(k+1) = A(k)x(k) + B(k)u(k), \qquad (5.32)$$

A(k), B(k) bounded above, is said to be *uniformly reachable*, if an $\alpha > 0$ and a finite integer $N > 0$ exist, such that

$$W(k,k-N) > \alpha\, I \qquad (5.33)$$

holds for all $k \in K$.

Therefore, if a system with the state equation (5.32) is uniformly reachable, the minimum amount of energy required for transferring the state from $x_o = x(k-N)$ to $x_f = x(k)$ does not depend on the time k, but it depends only on the duration N and the 'difference state' $\bar{x} = x_f - \Phi(k,k-N)x_o$.

In [16] a stabilizing control law

$$u(k) = -L(k)x(k), \qquad (5.34)$$

is constructed according to the following lemma:

5-6   LEMMA: With the system of order n with m inputs

$$x(k+1) = A(k)x(k) + B(k)u(k) \qquad (5.35)$$

uniformly reachable, A(k), B(k) bounded above, and the feedback gain m×n-matrix

$$L(k) := B^T(k)\Phi^T(k+N,k+1)W^{-1}(k+N,k)\Phi(k+N,k+1)A(k), \qquad (5.36)$$

the closed loop system

$$x(k+1) = [A(k)-B(k)L(k)]x(k) \qquad (5.37)$$

is exponentially stable.

EXAMPLE 5.1: Given the system $\{A(k),B(k)\}$ of order n=2 with the state transition equation

$$x(k+1) = \begin{bmatrix} 1 & a(k) \\ 0 & 1 \end{bmatrix} x(k) + \begin{bmatrix} 0 \\ 1 \end{bmatrix} u(k), \qquad (5.38)$$

and $a(k)\neq 0$ bounded above for $k\in K$. Since

$$\text{rank } W(k,k-1) = \text{rank} \begin{bmatrix} 0 & 0 \\ 0 & 1 \end{bmatrix} = 1 < n = 2, \qquad (5.39)$$

and

$$\text{rank } W(k,k-2) = \text{rank} \begin{bmatrix} a^2(k-1) & a(k-1) \\ a(k-1) & 2 \end{bmatrix} = n = 2, \qquad (5.40)$$

it is N=2. With

$$W^{-1}(k+2,k) = \begin{bmatrix} 2/a^2(k+1) & -1/a(k+1) \\ -1/a(k+1) & 1 \end{bmatrix} \qquad (5.41)$$

the feedback 1×2-matrix is

$$L(k) = B^T A^T(k+1)W^{-1}(k+2,k)A(k+1)A(k)$$

$$= [1/a(k+1), 1+a(k)/a(k+1)], \qquad (5.42)$$

which stabilizes the system. Indeed, the closed-loop system

$$x(k+1) = [A(k)-BL(k)]x(k)$$

$$= \begin{bmatrix} 1 & a(k) \\ -1/a(k+1) & -a(k)/a(k+1) \end{bmatrix} x(k) \qquad (5.43)$$

has the 'dead beat'-property, i.e., every initial state x_0 is steered to the zero state in utmost n=2 steps, because the state transition matrix $\bar{\Phi}(k_0+2,k_0)$ of the closed-loop system is the zero matrix,

$$\bar{\Phi}(k_o+2,k_o) = [A(k_o+1)-BL(k_o+1)] \cdot [A(k_o)-BL(k_o)] = 0, \quad (5.44)$$

and

$$x(k_o+2) = \bar{\Phi}(k_o+2,k_o)x(k_o) = 0 \qquad (5.45)$$

for all $x_o \in \mathbb{R}^2$. ∎

For linear *time-invariant* discrete-time systems of order n with the state transition equation

$$x(k+1) = A\,x(k) + B\,x(k), \qquad (5.46)$$

the *reachability matrix* is

$$S = [B,\ AB,\ A^2 B,\ \ldots,\ A^{n-1}B], \qquad (5.47)$$

the *reachability gramian*

$$W = \sum_{i=o}^{n-1} A^i B\ B^T A^{T^i}, \qquad (5.48)$$

and the *stabilizing feedback matrix* according to Lemma 5-6

$$L = B^T (A^T)^{n-1} W^{-1} A^n. \qquad (5.49)$$

If a time-invariant system is reachable, it is, of course, uniformly reachable too.

5.2 Observability

In control engineering it is often too expensive to measure
all state variables for feedback purpose, so only certain
output variables, combined in the *output vector* y(k), are
available. Usually these are some linear combinations of the
state variables, as y(k)=C(k)x(k), where $y(k) \in \mathbb{R}^r$, $C(k) \in \mathbb{C}^{r \times n}$,
and r<n. Such a linear time-variant discrete-time system
{A(k),B(k),C(k)} can be described by its state equation and
its output equation:

$$x(k+1) = A(k)x(k) + B(k)u(k), \qquad\qquad (5.50)$$

$$y(k) \quad = C(k)x(k). \qquad\qquad (5.51)$$

As mentioned above, measurements are usually available
only of the output vector y(k) and the problem is in computing
the state vector. But first the problem is to determine,
whether it is possible to compute the state, before setting
up a scheme to do it. If the *observability* of a system is
examined, it will be checked, whether the state x(k) can be
determined from future measurement data.

<table>
<tr><td>5-7</td><td>DEFINITION: A discrete-time system is said to be *observable at k*, if a finite integer N>0 exists so that any x=x(k) can uniquely be determined from the output sequence y(k), y(k+1),..., y(k+N-1) and the input sequence u(k), u(k+1),..., u(k+N-2).</td></tr>
</table>

A necessary and sufficient condition for the observability
of a system {A(k),B(k),C(k)} is:

$\underline{5-8}$

> LEMMA: The system $\{A(k),B(k),C(k)\}$ of order n is observable at k, if and only if for a finite N>0
>
> $$\text{rank } R(N,k) = n, \tag{5.52}$$
>
> where
>
> $$R(N,k) := \begin{bmatrix} C(k) \\ C(k+1)\Phi(k+1,k) \\ \vdots \\ C(k+N-1)\Phi(k+N-1,k) \end{bmatrix} \tag{5.53}$$
>
> is the so called *observability matrix*.

For a proof of Lemma 5-8 see e.g. [12].

If the observability $r \cdot N \times n$-matrix $R(N,k)$ has rank equal to n, the *observability gramian*

$$M(k+N,k) := R^T(N,k)R(N,k)$$

$$= \sum_{i=k}^{k+N-1} \Phi^T(i,k)C^T(i)C(i)\Phi(i,k) \tag{5.54}$$

also has rank equal to n, is symmetric, and positive definite, so one has:

$\underline{5-9}$

> LEMMA: The system $\{A(k),B(k),C(k)\}$ of order n is observable in k, if and only if for a finite N>0 the symmetric observability gramian
>
> $$M(k+N,k) = \sum_{i=k}^{k+N-1} \Phi^T(i,k)C^T(i)C(i)\Phi(i,k)$$
>
> is positive definite: $M(k+N,k)>0.$ $\tag{5.55}$

To compute the state vector, it is first easy to see, that

$$y(i) = C(i)\Phi(i,k)x(k) + \sum_{j=k}^{i-1} C(i)\Phi(i,j+1)B(j)u(j). \qquad (5.56)$$

With the definition

$$\bar{y}(i) := y(i) - \sum_{j=k}^{i-1} C(i)\Phi(i,j+1)B(j)u(j) \qquad (5.57)$$

one gets for $i=k,k+1,\ldots,k+N-1$ the sequence

$$\begin{aligned}
\bar{y}(k) \quad &= C(k)x(k) \\
\bar{y}(k+1) \quad &= C(k+1)\Phi(k+1,k)x(k) \\
&\;\;\vdots \\
\bar{y}(k+N-1) &= C(k+N-1)\Phi(k+N-1)x(k).
\end{aligned} \qquad (5.58)$$

The sequence (5.58) can be combined to

$$y(N,k) = R(N,k)x(k), \qquad (5.59)$$

whereby

$$y(N,k) := \begin{bmatrix} \bar{y}(k) \\ \bar{y}(k+1) \\ \vdots \\ \bar{y}(k+N-1) \end{bmatrix} \qquad (5.60)$$

Equation (5.59) from the left multiplied by the transposed observability matrix $R^T(N,k)$ yields

$$\begin{aligned}
R^T(N,k)y(N,k) &= R^T(N,k)R(N,k)x(k) \\
&= M(N,k)x(k), \qquad (5.61)
\end{aligned}$$

and finally the desired state vector

$$x(k) = M^{-1}(N,k) R^T(N,k) y(N,k),\qquad(5.62)$$

where the inverse matrix $M^{-1}(N,k)$ exists, if the system is observable.

A stronger form of observability is given in:

<table>
<tr><td>5-10</td><td>DEFINITION: The system $\{A(k),B(k),C(k)\}$, $A(k),B(k)$ bounded above, is said to be uniformly observable, if a finite $\alpha>0$ and a finite $N>0$ exist such that

$$M(k+N,k) > \alpha I \qquad(5.63)$$

holds for all $k\in K$.</td></tr>
</table>

Now a scheme is given for reconstructing the state of an uniformly observable system from observations of its input and output values. The state can be computed by an *observer*; this is itself a dynamical system (<u>Fig.5/1</u>). Its inputs are

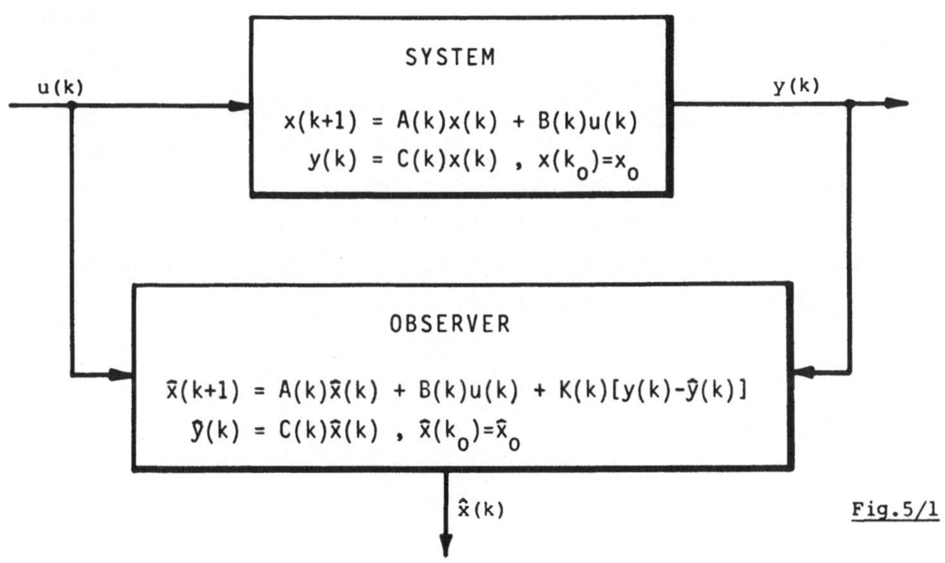

<u>Fig.5/1</u>

the values of the inputs u and the measured outputs y from
the system, and its state vector $\hat{x}$ generates the state x of
the original system. The observer is given by the equations

$$\hat{x}(k+1) = A(k)\hat{x}(k) + B(k)u(k) + K(k)[y(k)-\hat{y}(k)], \qquad (5.64)$$

$$\hat{y}(k) \quad = C(k)\hat{x}(k). \qquad (5.65)$$

It is easy to see, that the *observation error vector*

$$\tilde{x}(k) := x(k) - \hat{x}(k) \qquad (5.66)$$

obeys the difference equation

$$\tilde{x}(k+1) = [A(k)-K(k)C(k)]\tilde{x}(k). \qquad (5.67)$$

If the 'error system' (5.67) is asymptotically or exponentially stable, then

$$\lim_{k\to\infty} \tilde{x}(k) = 0, \qquad (5.68)$$

i.e.,

$$\lim_{k\to\infty} \hat{x}(k) = x(k). \qquad (5.69)$$

If the system matrix of the error system (5.67) is transposed, then it is $A^T(k)-C^T(k)K^T(k)$. Comparing this system
matrix with the system matrix $A(k)-B(k)L(k)$ in (5.37), one
has following dual situation: given $A^T(k)$ and $C^T(k)$, a matrix
$K^T(k)$ is searched, such that $A^T(k)-C^T(k)K^T(k)$ is exponentially stable. With the dualities

$$A^T \triangleq A$$
$$C^T \triangleq B \qquad (5.70)$$
$$K^T \triangleq L$$

from (5.36) one has

$$K^T(.) = C(.)\Phi(.,.)M^{-1}(.,.)\Phi^T(.,.)A^T(.). \tag{5.71}$$

By a carefully analysis of the duality between the control problem and the observer problem, in [16] these duality relationships are derived:

$$A(k) \triangleq A^T(-k),$$

$$C(k) \triangleq B^T(-k), \tag{5.72}$$

and from that

$$M(i,k) = W(-k,-i) \quad \text{for } i>k, \tag{5.73}$$

$$K(k) = L^T(-k). \tag{5.74}$$

So the following lemma is obtained:

5-11   LEMMA: With the system $\{A(k),B(k),C(k)\}$ of order n, m inputs, and r outputs, uniformly observable, $A(k),C(k)$ bounded above, and the observer n×r-matrix

$$K(k)=A(k)\Phi(k,k-N+1)M^{-1}(k+1,k-N+1)\Phi^T(k,k-N+1)C^T(k) \tag{5.75}$$

the error system

$$\tilde{x}(k+1) = [A(k)-K(k)C(k)]\tilde{x}(k) \tag{5.76}$$

is exponentially stable.

EXAMPLE 5.2: Given the system in Example 5.1 with the state equation (5.38) and the output equation

$$y(k) = [1,0]x(k). \tag{5.77}$$

For this system is

$$\text{rank } M(k+1,k) = \text{rank } \begin{bmatrix} 1 & 0 \\ 0 & 0 \end{bmatrix} = 1 < n = 2,$$

and

$$\text{rank } M(k+2,k) = \text{rank } \begin{bmatrix} 2 & a(k) \\ a(k) & a^2(k) \end{bmatrix} = n = 2,$$

i.e., N=2. With

$$M^{-1}(k+1,k-N+1) = M^{-1}(k+1,k-1)$$

$$= \begin{bmatrix} 1 & -1/a(k-1) \\ -1/a(k-1) & 2/a^2(k-1) \end{bmatrix} \qquad (5.78)$$

the observer 2×1-matrix is

$$K(k) = A(k)A(k-1)M^{-1}(k+1,k-1)A^T(k-1)C^T$$

$$= \begin{bmatrix} 1+a(k)/a(k-1) \\ 1/a(k-1) \end{bmatrix}. \qquad (5.79)$$

The error system

$$\widetilde{x}(k+1) = [A(k)-K(k)C]\widetilde{x}(k)$$

$$= \begin{bmatrix} -a(k)/a(k-1) & a(k) \\ -1/a(k-1) & 1 \end{bmatrix} \widetilde{x}(k) \qquad (5.80)$$

has again, as the closed-loop system in Example 5.1, the
'dead beat'-property: every initial error vector $\widetilde{x}(k_o)$ is
transferred to the zero vector in utmost n=2 steps, because

$$\widetilde{x}(k_o+2) = [A(k_o+1)-K(k_o+1)C]\cdot[A(k_o)-K(k_o)C]\widetilde{x}(k_o)$$

$$= 0\cdot\widetilde{x}(k_o) = 0, \qquad (5.81)$$

i.e.,

$$\hat{x}(k_o+2) = x(k_o+2),$$

the observer state $\hat{x}$ is equal to the system state x after utmost two steps. ∎

For linear *time-invariant* discrete-time systems of order n with the mathematical description

$$x(k+1) = A\ x(k) + B\ u(k), \tag{5.82}$$

$$y(k) = C\ x(k), \tag{5.83}$$

the *observability matrix* is

$$R = \begin{bmatrix} C \\ CA \\ \vdots \\ CA^{n-1} \end{bmatrix}, \tag{5.84}$$

the *observability gramian*

$$M = \sum_{i=0}^{n-1} (A^T)^i C^T C A^i, \tag{5.85}$$

and the *observer matrix*, according to Lemma 5-11,

$$K = A^n M^{-1} (A^T)^{n-1} C^T. \tag{5.86}$$

The observability of a time-invariant system implies uniform observability.

5.3 STABILIZABILITY

This and the following sections are closely related to [1].
If a state x converges only slowly, or even diverge, i.e., if

$$\|\Phi(k,k-t)x\| \geq \beta\|x\| \quad , \quad \beta\in[0,1], \tag{5.87}$$

then this state x should be transferred to the zero state
(be stabilized) by a measurable amount of 'input energy'.
But by the explanations of the last section this is only
possible, if

$$x^T W(k,k-t)x \geq \alpha \ x^T x, \quad \alpha\in(0,\infty). \tag{5.88}$$

For those states, which converge sufficiently quick to zero,
the inequality (5.88) must not be valid! So one has the
following definition:

5-12   DEFINITION: The system $\{A(k),B(k),C(k)\}$ is said to be
uniformly stabilizable, if there exist
integers $s\geq t\geq 0$ and constants α, β with
$\alpha\in(0,\infty)$, $\beta\in[0,1]$, such that, whenever

$$\|\Phi(k,k-t)x\| \geq \beta\|x\| \tag{5.89}$$

for some x, k, then

$$x^T W(k,k-s)x \geq \alpha \ x^T x, \tag{5.90}$$

where

$$W(k,k-s) = \sum_{i=k-s}^{k-1} \Phi(k,i+1)B(i)B^T(i)\Phi^T(k,i+1). \tag{5.91}$$

A comparison of this definition with Definition 5-5 of
uniform reachability leads at once to the following lemma:

5-13   **LEMMA:** If the system $\{A(k),B(k),C(k)\}$ is uniformly
reachable, then it is uniformly stabilizable.

The opposite direction of Lemma 5-13 naturally is not true.

Now an input sequence is constructed which transfers any
initial state of an only uniformly stabilizable system to the
zero state. First this lemma is needed:

5-14   **LEMMA:** Let $T(k)$ be an orthogonal matrix, $T^{-1}(k)=T^{T}(k)$,
and suppose that state equation for $\bar{x}(k)=T(k)x(k)$
is constructed from

$$x(k+1) = A(k)x(k) + B(k)u(k), \qquad (5.92)$$

i.e.,

$$\bar{x}(k+1) = T(k+1)A(k)T^{T}(k)\bar{x}(k) + T(k+1)B(k)u(k)$$

$$= \bar{A}(k)\bar{x}(k) + \bar{B}(k)u(k). \qquad (5.93)$$

Then

$$\bar{W}(k,k-s) = T(k)W(k,k-s)T^{T}(k). \qquad (5.94)$$

PROOF: It is

$$T(k)W(k,k-s)T^{T}(k) = \sum_{i=k-s}^{k-1} T(k)\Phi(k,i+1)B(i)B^{T}(i)\Phi^{T}(k,i+1)T^{T}(k)$$

$$= \sum_{i=k-s}^{k-1} T(k)\Phi(k,i+1)T^{T}(i+1)T(i+1)B(i)B^{T}(i)T^{T}(i+1)T(i+1)\Phi^{T}(k,i+1)\times$$
$$\times T^{T}(k) =$$

$$= \sum_{i=k-s}^{k-1} \bar{\Phi}(k,i+1)\bar{B}(i)\bar{B}^T(i)\bar{\Phi}^T(k,i+1)$$

$$= \bar{W}(k,k-s).\blacksquare \tag{5.95}$$

Now a transformation matrix $T(k)$ is selected such that $\bar{W}(k,k-s)$ is diagonal, with diagonal entries ordered in decreasing magnitude. Suppose this is done, and the bar is omitted in that what follows. For an uniformly stabilizable system is written

$$W(k,k-s) = \begin{bmatrix} W_1(k,k-s) & O \\ O & W_2(k,k-s) \end{bmatrix}, \tag{5.96}$$

where the diagonal entries of $W_1(k,k-s)$ are all greater than or equal to β, and those of $W_2(k,k-s)$ are all less than β. The dimension of $W_1(k,k-s)$ may not be constant with k. Then one obtains

<u>5-15</u>   LEMMA: Suppose an uniformly stabilizable system $\{A(k),B(k),C(k)\}$. Let $W_1(k,k-s)$ be as defined above. Let the initial state $x=x(k-s)$ be arbitrary. Define an input sequence $u(.)$ by

$$u(i) = \begin{cases} -B^T(i)\Phi^T(k,i+1)\begin{bmatrix} W_1^{-1}(k,k-s) & O \\ O & O \end{bmatrix}\Phi(k,k-s)x \\ \qquad\qquad\qquad\qquad\qquad \text{for } i\in[k-s,k-1], \\ O \qquad \text{for } i \geq k. \end{cases} \tag{5.97}$$

Then for $i \geq 0$

$$x(k+i) = \Phi(k+i,k)\begin{bmatrix} O & O \\ O & I \end{bmatrix}\Phi(k,k-s)x. \tag{5.98}$$

The partitioning of the matrix $\begin{bmatrix} O & O \\ O & I \end{bmatrix}$ in

$$\boxed{(5.98) \text{ is like that of } W(k,k-s) \text{ in } (5.96).}$$

PROOF: The input sequence of (5.97) may be combined as

$$u(k,s) := \begin{bmatrix} u(k-1) \\ u(k-2) \\ \vdots \\ u(k-s) \end{bmatrix}$$

$$= - \begin{bmatrix} B^T(k-1) \\ B^T(k-2)\Phi^T(k,k-1) \\ \vdots \\ B^T(k-s)\Phi^T(k,k-s+1) \end{bmatrix} \begin{bmatrix} W_1^{-1}(k,k-s) & O \\ & \\ O & O \end{bmatrix} \Phi(k,k-s)x$$

$$= -S^T(k,k-s) \begin{bmatrix} W_1^{-1}(k,k-s) & O \\ & \\ O & O \end{bmatrix} \Phi(k,k-s)x. \qquad (5.99)$$

Additional it is

$$x(k) = \Phi(k,k-s)x + \sum_{i=k-s}^{k-1} \Phi(k,i+1)B(i)u(i)$$

$$= \Phi(k,k-s)x + S(k,k-s)u(k,s)$$

$$= \Phi(k,k-s)x - S(k,k-s)S^T(k,k-s) \begin{bmatrix} W_1^{-1}(k,k-s) & O \\ & \\ O & O \end{bmatrix} \Phi(k,k-s)x$$

$$= \Phi(k,k-s)x - \begin{bmatrix} W_1(k,k-s) & O \\ & \\ O & W_2(k,k-s) \end{bmatrix} \begin{bmatrix} W_1^{-1}(k,k-s) & O \\ & \\ O & O \end{bmatrix} \Phi(k,k-s)x$$

$$= \begin{bmatrix} O & O \\ & \\ O & I \end{bmatrix} \Phi(k,k-s)x. \qquad (5.100)$$

For $u(k+i)\equiv 0$, $i\geq 0$, finally one receives for the 'initial state' (5.100):

$$x(k+i) \;=\; \Phi(k+i,k)\begin{bmatrix} O & O \\ O & I \end{bmatrix}\Phi(k,k-s)x.\;\blacksquare$$

In case uniform reachability is present, one has $x(k)=O$. If it is lacking, (5.98) holds, this ensures that the state asymptotically is zero; because from the stabilizability Definition 5-12 is following that any vector of the form $\begin{bmatrix} O \\ x_2 \end{bmatrix}$ has the property

$$\left\| \Phi(k+t,k)\begin{bmatrix} O \\ x_2 \end{bmatrix} \right\| \;<\; \beta\|x_2\| \;,$$

or for *all* x_1, x_2 of appropriate dimension

$$\left\| \Phi(k+t,k)\begin{bmatrix} O & O \\ O & I \end{bmatrix}\begin{bmatrix} x_1 \\ x_2 \end{bmatrix} \right\| \;<\; \beta\left\| \begin{bmatrix} x_1 \\ x_2 \end{bmatrix} \right\| \;,$$

so that

$$\left\| \Phi(k+t,k)\begin{bmatrix} O & O \\ O & I \end{bmatrix} \right\| \;<\; \beta \;<\; 1 ,$$

or

$$\lim_{k\to\infty}\left\| \begin{bmatrix} x_1 \\ x_2 \end{bmatrix} \right\| \;=\; O .$$

5.4 DETECTABILITY

For *time-invariant* systems the following detectability definition has been given [10]: a linear time-invariant discrete-time system is *detectable*, if its unreconstructable subspace is contained within its stable subspace; which requires that the unstable modes of the system are observable.

Now a time-varying version of this notion is given so,
that when a state trajectory is not fast decaying, then that
trajectory must be observable.

<u>5-16</u> DEFINITION: The system $\{A(k),B(k),C(k)\}$ is said to be
uniformly detectable, if there exist
integers $s \geq t \geq 0$ and constants α, β with
$\alpha \in (0,\infty)$, $\beta \in [0,1)$ such that, whenever

$$\|\Phi(k+t,k)x\| \geq \beta\|x\| \qquad (5.101)$$

for some x and k, then

$$x^T M(k+s,k)x \geq \alpha\ x^T x, \qquad (5.102)$$

where

$$M(k+s,k) := \sum_{i=k}^{k+s-1} \Phi^T(i,k)C^T(i)C(i)\Phi(i,k). \qquad (5.103)$$

Since $M(k+s,k) > \alpha I$ is a sufficient condition for detectability as above, a comparison with the uniform observability
condition of Definition 5-10 immediately leads to:

<u>5-17</u> LEMMA: If the system $\{A(k),B(k),C(k)\}$ is uniformly
observable, then it is uniformly detectable.

Now a state observer is constructed which reconstructs an
estimate $\hat{x}(k)$ of the system state $x(k)$ of an only uniformly
detectable system. First again a partition of the reachability gramian is needed and first:

5-18   LEMMA: Let $T(k)$ be an orthogonal matrix and suppose that state variable equation for $\bar{x}(k)=T(k)x(k)$ are constructed from

$$x(k+1) = A(k)x(k) + B(k)u(k), \qquad (5.104)$$

$$y(k) = C(k)x(k), \qquad (5.105)$$

that is

$$\bar{x}(k+1) = T(k+1)A(k)T^T(k)\bar{x}(k)+T(k+1)B(k)u(k)$$

$$=: \bar{A}(k)\bar{x}(k) + \bar{B}(k)u(k), \qquad (5.106)$$

$$y(k) = C(k)T^T(k)\bar{x}(k)$$

$$=: \bar{C}(k)\bar{x}(k). \qquad (5.107)$$

Then

$$\bar{M}(k+s,k) = T(k)M(k+s,k)T^T(k). \qquad (5.108)$$

PROOF:

$$T(k)M(k+s,k)T^T(k) = \sum_{i=k}^{k+s-1} T(k)\Phi^T(i,k)C^T(i)C(i)\Phi(i,k)T^T(k)$$

$$= \sum_{i=k}^{k+s-1} [T(i)\Phi(i,k)T^T(k)]^T[C(i)T^T(i)]^TC(i)T^T(i)T(i)\Phi(i,k)T^T(k)$$

$$= \sum_{i=k}^{k+s-1} \bar{\Phi}^T(i,k)\bar{C}^T(i)\bar{C}(i)\bar{\Phi}(i,k)$$

$$= \bar{M}(k+s,k). \blacksquare \qquad (5.109)$$

As in the preceding section a transformation matrix $T(k)$ is selected such that $\bar{M}(k+s,k)$ is diagonal, with diagonal entries ordered in decreasing magnitude. Suppose this is done and the bar is omitted in the following. Also for an uniformly detectable system is written

$$M(k+s,k) = \begin{bmatrix} M_1(k+s,k) & O \\ O & M_2(k+s,k) \end{bmatrix}, \tag{5.110}$$

where the diagonal entries of $M_1(k+s,k)$ are all greater than or equal to β, and those of $M_2(k+s,k)$ are all less than β. The dimension of $M_1(k+s,k)$ may vary with k. Then it is [1]:

5-19 LEMMA: Suppose the system $\{A(k),B(k),C(k)\}$ is uniformly detectable, and let

$$G(k,j) := C(k)\Phi(k,j+1)B(j) \tag{5.111}$$

denote the impulse response matrix of this system. Let k_o be an arbitrary but fixed integer, and let $M_1(k+s,k)$ be as defined above. Let the input sequence $u(.)$ and $x(k_o-t+1)$ be arbitrary. Define a sequence $\hat{x}(.)$

$$\hat{x}(k) = O \quad \text{for} \quad k\in[k_o-t+1,k_o], \tag{5.112}$$

and for $k \geq k_o-t+1$ by

$$\hat{x}(k+t) = \Phi(k+t,k)\left\{ \begin{bmatrix} M_1^{-1}(k+s,k) & O \\ O & O \end{bmatrix} \sum_{i=k}^{k+s-1} \Phi^T(i,k)C^T(i) \times \right.$$

$$\times \left[y(i) - \sum_{j=k}^{i-1} G(i,j)u(j) \right] + \begin{bmatrix} O & O \\ O & I \end{bmatrix} \hat{x}(k) \Bigg\} +$$

$$+ \sum_{j=k}^{k+t-1} \Phi(k+t,j+1)B(j)u(j). \tag{5.113}$$

Then for $k \geq k_0 - t + 1$

$$\hat{x}(k+t) - x(k+t) = \Phi(k+t,k) \begin{bmatrix} O & O \\ O & I \end{bmatrix} [\hat{x}(k) - x(k)].$$

$$(5.114)$$

PROOF: Observe that

$$y(i) = C(i)x(i)$$

$$= C(i)\Phi(i,k)x(k) + \sum_{j=k}^{i-1} G(i,j)u(j)$$

or

$$y(i) - \sum_{j=k}^{i-1} G(i,j)u(j) = C(i)\Phi(i,k)x(k). \qquad (5.115)$$

(5.115) in (5.113) gives, with the definition of $M(k+s,k)$,

$$\hat{x}(k+t) = \Phi(k+t,k)\left\{ \begin{bmatrix} M_1^{-1}(k+s,k) & O \\ O & O \end{bmatrix} \sum_{i=k}^{k+s-1} \Phi^T(i,k)C^T(i)C(i)\Phi(i,k)x(k) + \right.$$

$$\left. + \begin{bmatrix} O & O \\ O & I \end{bmatrix} \hat{x}(k) \right\} + \sum_{j=k}^{k+t-1} \Phi(k+t,j+1)B(j)u(j) =$$

$$= \Phi(k+t,k)\left\{ \begin{bmatrix} M_1^{-1}(k+s,k) & O \\ O & O \end{bmatrix} \begin{bmatrix} M_1(k+s,k) & O \\ O & M_2(k+s,k) \end{bmatrix} x(k) + \right.$$

$$\left. + \begin{bmatrix} O & O \\ O & I \end{bmatrix} \hat{x}(k) \right\} + \sum_{j=k}^{k+t-1} \Phi(k+t,j+1)B(j)u(j) =$$

$$
=\Phi(k+t,k)\left\{\begin{bmatrix} I & 0 \\ 0 & 0 \end{bmatrix} x(k) + \begin{bmatrix} 0 & 0 \\ 0 & I \end{bmatrix} \hat{x}(k)\right\} +
$$

$$
+ \sum_{j=k}^{k+t-1} \Phi(k+t,j+1)B(j)u(j)
$$

$$
= \Phi(k+t,k)x(k) + \sum_{j=k}^{k+t-1} \Phi(k+t,j+1)B(j)u(j) +
$$

$$
+ \Phi(k+t,k) \begin{bmatrix} 0 & 0 \\ 0 & I \end{bmatrix} [\hat{x}(k)-x(k)]
$$

$$
= x(k+t) + \Phi(k+t,k) \begin{bmatrix} 0 & 0 \\ 0 & I \end{bmatrix} [\hat{x}(k)-x(k)]. \blacksquare \tag{5.116}
$$

If the system is uniformly observable, one has $\hat{x}(k+t)=x(k+t)$. If the system is only uniformly detectable, i.e. (5.114) is holding, and this ensures that the observation error asymptotically is zero, as argued below.

5.5 Relations between ℓ^p-BIBO- and ℓ^p-BIBS-Stability

After this long take-off run across section 5.1 to section 5.4 the main results of this chapter are obtained [1]: given an uniformly detectable system, then ℓ^p-BIBO-stability implies ℓ^p-BIBS-stability. The opposite direction of this statement is for bounded $C(k)$ obviously true:

<table>
<tr><td>5-20</td><td>THEOREM: Consider the system$\{A(k),B(k),C(k)\}$ with $A(k)$, $B(k)$, $C(k)$ resp. bounded. Then ℓ^p-BIBS-stability implies ℓ^p-BIBO-stability.</td></tr>
</table>

From Lemma 5-18 it is clear that there is no loss of
generality in establishing that ℓ^p-BIBO-stability implies
ℓ^p-BIBS-stability for a linear time-variant discrete-time
system for which for each k, M(k+s,k) is diagonal, with
diagonal entries ordered in decreasing magnitude.

<u>5-21</u>   THEOREM: Consider the system {A(k),B(k),C(k)} with
bounded matrices A(k), B(k), C(k) resp.
Then for anyone $p\in[1,\infty]$, if the system is
ℓ^p-BIBO-stable and uniformly detectable,
then it is ℓ^p-BIBS-stable.

PROOF: Assume that M(k+s,k) is diagonal with M_1(k+s,k) as
described above in Lemma 5-19. Equation (5.113) can
be reorganized as

$$\hat{x}(k+t) = \Phi(k+t,k) \begin{bmatrix} O & O \\ O & I \end{bmatrix} \hat{x}(k) + w(k), \qquad (5.117)$$

where w(k) is a finite length moving average of u(.)
and y(.) terms, with bounded weights. The ℓ^p-BIBO-
stability hypothesis implies that if

$$\left[\sum_{k \geq k_o} \|u(k)\|^p \right]^{1/p} \leq c, \qquad (5.118)$$

then

$$\left[\sum_{k \geq k_o} \|w(k)\|^p \right]^{1/p} \leq \eta \cdot c \qquad (5.119)$$

for some $\eta > 0$, independent of c, u(.), and k_o. also,
recalling the special form of M(k+s,k) in (5.110),
one sees from the detectability definitions that any
vector of the form $\begin{bmatrix} O \\ x_2 \end{bmatrix}$ has the property

$$\left\|\Phi(k+t,k)\begin{bmatrix} O \\ x_2 \end{bmatrix}\right\| < \beta \, \|x_2\|, \tag{5.120}$$

or for *all* x_1, x_2 of appropriate dimension

$$\left\|\Phi(k+t,k)\begin{bmatrix} O & O \\ O & I \end{bmatrix}\begin{bmatrix} x_1 \\ x_2 \end{bmatrix}\right\| < \beta \left\|\begin{bmatrix} x_1 \\ x_2 \end{bmatrix}\right\|, \tag{5.121}$$

so that

$$\left\|\Phi(k+t,k)\begin{bmatrix} O & O \\ O & I \end{bmatrix}\right\| < \beta < 1. \tag{5.122}$$

This means, that (5.117) is ℓ^p-BIBO-stable by Lemma 5-19. Using the $w(.)$ bound, one sees that $\hat{x}(k)$ is such that

$$\left[\sum_{k \geq k_o} \|\hat{x}(k)\|^p\right]^{1/p} < \delta \cdot c \tag{5.123}$$

for some δ, independent of c, $u(.)$, and k_o. The boundary conditions on $\hat{x}(k)$ in Lemma 5-19 and on $x(k)$ in the BIBO-stability definition imply $\hat{x}(k)-x(k)=O$ for $k=k_o-t+1,k_o-t+2,\ldots,k_o$ and so by (5.114), $\hat{x}(k)=x(k)$ for all k. The result then is following from (5.123).∎

The statements of Theorem 5-20 and Theorem 5-21 are visualized in <u>Fig.5/2</u>.

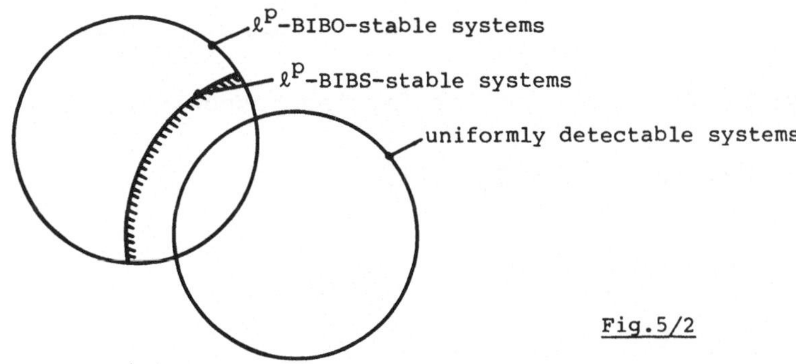

<u>Fig.5/2</u>

5.6 Relations between ℓ^P-BIBO-Stability and Exponential Stability

In Theorem 5-20 was stated, that ℓ^P-BIBO-stability and uniform detectability imply ℓ^P-BIBS-stability. Now it shall be shown, that ℓ^P-BIBS-stability and uniform stabilizability imply exponential stability of the homogeneous state equation of a linear time-variant discrete-time system. First two lemmata must be stated.

<u>5-22</u> **Lemma:** Let $G(k,j)$ be the impulse response matrix of the linear time-variant discrete-time system with the mathematical description

$$x(k+1) = A(k)x(k) + B(k)u(k) \qquad (5.124)$$

$$y(k) = C(k)x(k) \qquad (5.125)$$

and $\bar{G}(k,j)$ the impulse response matrix of the *adjoint system*

$$\bar{x}(k+1) = A^T(-k)\bar{x}(k) + \bar{C}(-k)\bar{u}(k) \qquad (5.126)$$

$$\bar{y}(k) = B^T(-k)\bar{x}(k). \qquad (5.127)$$

Then

$$\bar{G}(k,j) = G^T(-j,-k). \qquad (5.128)$$

Proof: By definition of the impulse response matrix it is

$$\bar{G}(k,j) = B^T(-k)\bar{\Phi}(k,j+1)C^T(-j), \qquad (5.129)$$

and with

$$\bar{\Phi}(k,j+1) = A^T(-(k-1))A^T(-(k-2))\ldots A^T(-(j+1)) =$$

$$= [A(-j-1)\ldots A(-k+2)A(-k+1)]^T$$

$$= \Phi^T(-j,-k+1) \tag{5.130}$$

it is

$$\bar{G}(k,j) = [C(-j)\Phi(-j,-k+1)B(-k)]^T$$

$$= G^T(-j,-k). \blacksquare \tag{5.131}$$

<u>5-23</u>

> **LEMMA:** The system $\{A(k),B(k),C(k)\}$ is ℓ^p-BIBO-stable for $p \in [1,\infty]$ if and only if the adjoint system with the mathematical description (5.126/5.127) is ℓ^q-BIBO-stable for q satisfying $q^{-1}+p^{-1}=1$.

PROOF: Using the HÖLDER inequality one obtains the following set of equivalences:

$\{A(k),B(k),C(k)\}$ is ℓ^p-BIBO-stable $\leftrightarrow$

$$\leftrightarrow \; \forall_{u(.)\in\ell^p} \; \forall_{\bar{u}(.)\in\ell^q} \left\| \sum_{k=-\infty}^{+\infty} \bar{u}^T(k)y(k) \right\| = \left\| \sum_{k=-\infty}^{+\infty} u^T(k) \; \times \right.$$

$$\left. \times \sum_{j=-\infty}^{k-1} G(k,j)u(j) \right\| \leq c \left[\sum_{k=-\infty}^{+\infty} \|u(k)\|^p \right]^{\frac{1}{p}} \left[\sum_{k=-\infty}^{+\infty} \|\bar{u}(k)\|^q \right]^{\frac{1}{q}}$$

$$\leftrightarrow \left\| \sum_{j=-\infty}^{+\infty} \left[\sum_{k=j+1}^{\infty} u^T(k)G(k,j) \right] u(j) \right\| \leq c \left[\sum_{k=-\infty}^{+\infty} \|u(k)\|^p \right]^{\frac{1}{p}} \left[\sum_{k=-\infty}^{+\infty} \|\bar{u}(k)\|^q \right]^{\frac{1}{q}}$$

$$\leftrightarrow \left\| \sum_{m=-\infty}^{+\infty} u^T(-m) \left[\sum_{i=-\infty}^{m-1} G^T(-i,-m)\bar{u}(-i) \right] \right\| \leq$$

$$\leq c \left[\sum_{m=-\infty}^{+\infty} \|u(-m)\|^p \right]^{1/p} \left[\sum_{i=-\infty}^{+\infty} \|\bar{u}(-i)\|^q \right]^{1/q}$$

$$\leftrightarrow \left\| \sum_{m=-\infty}^{+\infty} u^T(-m) \left[\sum_{i=-\infty}^{m-1} \bar{G}(m,i)\bar{u}(-i) \right] \right\| \leq$$

$$\leq c \left[\sum_{m=-\infty}^{+\infty} \| u(-m) \|^p \right]^{1/p} \left[\sum_{i=-\infty}^{+\infty} \| u(-i) \|^q \right]^{1/q}$$

$$\leftrightarrow \left\| \sum_{m=-\infty}^{+\infty} u^T(-m)\bar{y}(m) \right\| \leq \bar{c}$$

$\leftrightarrow$ the adjoint system is ℓ^q-BIBO-stable. $\blacksquare$

With Lemma 5-22 and Lemma 5-23 one obtains following main result:

<u>5-24</u> | THEOREM: Suppose the system $\{A(k),B(k),C(k)\}$ is uniformly stabilizable and uniformly detectable. Then $A(k)\in\mathcal{ES}$, if $\{A(k),B(k),C(k)\}$ is ℓ^p-BIBO-stable for anyone $p\in[1,\infty]$.

PROOF: If the system $\{A(k),B(k),C(k)\}$ is ℓ^p-BIBO-stable and uniformly detectable, then it is ℓ^p-BIBS-stable by Theorem 5-21. Then the system with the mathematical description

$$x(k+1) = A(k)x(k) + B(k)u(k) \tag{5.132}$$

$$y(k) = I\,x(k) \tag{5.133}$$

is ℓ^p-BIBO-stable and by Lemma 5-23 the adjoint system with the mathematical description

$$\bar{x}(k+1) = A^T(-k)\bar{x}(k) + \bar{u}(k) \tag{5.134}$$

$$\bar{y}(k) = B^T(-k)\bar{x}(k) \tag{5.135}$$

is ℓ^q-BIBO-stable. By Theorem 5-21, (5.134) is ℓ^q-BIBS-stable, so the system with the state equation (5.134) and the output equation

$$\bar{y}(k) = I \; \bar{x}(k) \tag{5.136}$$

is ℓ^q-BIBO-stable. But then by Lemma 5-23 the system

$$x(k+1) = A(k)x(k) + u(k) \tag{5.137}$$

$$y(k) = I \; x(k) \tag{5.138}$$

is ℓ^p-BIBO-stable, and in particular ℓ^p-BIBS-stable. By Theorem 4-9 (with $f(k)=u(k)$), the associated homogeneous equation is exponentially stable.∎

From Theorem 4-16 and Theorem 5-24 immediately follows:

<table>
<tr><td>5-25</td><td>THEOREM: Suppose, the system $\{A(k),B(k),C(k)\}$ is uniformly stabilizable and uniformly detectable. Then it is ℓ^p-BIBO-stable for anyone $p\in[1,\infty]$, if and only if it is ℓ^p-BIBO-stable for all $p\in[1,\infty]$.</td></tr>
</table>

For $y(k)=x(k)$, that is for $C(k)=I$, directly one obtains from Theorem 5-24:

<table>
<tr><td>5-26</td><td>THEOREM: If the system $\{A(k),B(k)\}$ is uniformly stabilizable and ℓ^p-BIBS-stable, then $A(k)\in\mathcal{ES}$.</td></tr>
</table>

PROOF: Since $C(k)=I$ implies uniform observability, and this implies uniform detectability, the hypothesis of Theorem 5-24 is valid and with $p=\infty$ Theorem 5-26 is following.∎

Since the system $\{A(k),B(k),C(k)\}$ is uniformly stabilizable and uniformly detectable if it is uniformly reachable
and uniformly observable, from Theorem 5-25 follows at once
the corollary:

<u>5-27</u> THEOREM: If the system $\{A(k),B(k),C(k)\}$ is uniformly
reachable, uniformly observable, and ℓ^P-BIBO-
stable, then $A(k)\in\mathcal{ES}$.

5.7 Summary of the Relations between Internal and External Stability

In the preceding sections of this and the foregoing
chapter for linear time-variant discrete-time systems with
the mathematical description

$$x(k+1) = A(k)x(k) + B(k)u(k)$$

$$y(k) = C(k)x(k)$$

with *bounded* matrices A(k), B(k), C(k), following relations
between internal and external stability were derived:

Theorem 4-16 and 5-20: $\mathcal{ES} \subset \ell^P\text{-BIBS-}\mathcal{S} \subset \ell^P\text{-BIBO-}\mathcal{S}$, (5.139)

Theorem 5-26 : $\ell^P\text{-BIBS-}\mathcal{S} \cap \mathcal{UR} \subset \mathcal{ES}$, (5.140)

Theorem 5-21 : $\ell^P\text{-BIBO-}\mathcal{S} \cap \mathcal{UO} \subset \ell^P\text{-BIBS-}\mathcal{S}$, (5.141)

Theorem 5-24 : $\ell^P\text{-BIBO-}\mathcal{S} \cap \mathcal{UR} \cap \mathcal{UO} \subset \mathcal{ES}$, (5.142)

whereby:

> DEFINITION: 1) ℓ^p-BIBS-$\mathcal{S}$:$\leftrightarrow$ set of all ℓ^p-BIBS-stable
>   systems $\{A(k),B(k)\}$. (5.143)
>
> 2) ℓ^p-BIBO-$\mathcal{S}$:$\leftrightarrow$ set of all ℓ^p-BIBO-stable
>   systems $\{A(k),B(k),C(k)\}$.
>   (5.144)
>
> 3) $\mathcal{US}_{\mathcal{z}}$:$\leftrightarrow$ set of all uniformly stabilizable
>   systems $\{A(k),B(k)\}$.   (5.145)
>
> 4) $\mathcal{UD}_{\mathcal{t}}$:$\leftrightarrow$ set of all uniformly detectable
>   systems $\{A(k),B(k),C(k)\}$. (5.146)

The relations (5.139) to (5.142) are visualized by <u>Fig.5/3</u>, where the horizontally hatched domain is the set ℓ^p-BIBO-$\mathcal{S}$ $\cap$ $\mathcal{US}_{\mathcal{z}}$ $\cap$ $\mathcal{UD}_{\mathcal{t}}$ of Theorem 5-24.

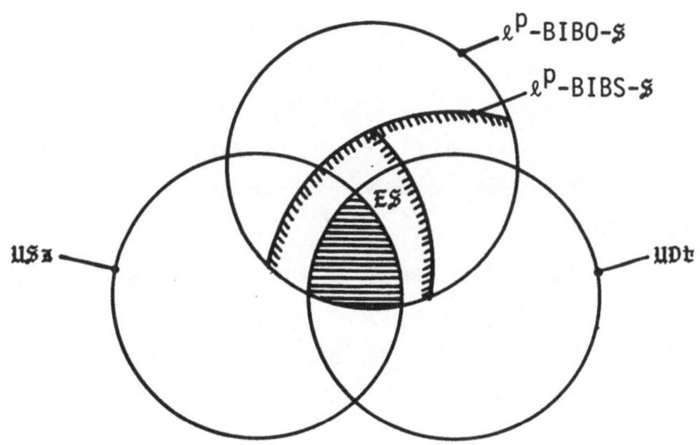

<u>Fig.5/3</u>

References

[1] Anderson,B.D.O.   Internal and External Stability of
                      Linear Time-Varying Systems.
                      SIAM J.Control a.Optim.20(1982)571-586

[2] Brockett,R.W.   Finite Dimensional Linear Systems.
                      New York: J.Wiley 1970

[3] Cesari,L.   Asymptotic Behavior and Stability
                      Problems in Ordinary Differential
                      Equations.
                      Berlin, Heidelberg, New York: Springer
                      1976

[4] Coddington,E.A.   Theory of Ordinary Differential Equations.
    Levinson,N.   New York: McGraw-Hill 1955

[5] Conti,R.   Linear Differential Equations and Control.
                      London, New York: Academic Press 1976

[6] D'Angelo,H.   Linear Time-Varying Systems: Analysis
                      and Synthesis.
                      Boston: Allyn and Bacon 1970

[7] Gantmacher,F.R.   Matrizenrechnung.
                      Berlin: Verlag der Wissenschaft 1970

[8] Hahn,W.   Stability of Motion.
                      Berlin, Heidelberg, New York: Springer
                      1965

[9] Kalman,R.E.   Control System Analysis and Design Via
    Bertram,J.E.   the "Second Method" of Lyapunov.
                      Trans. of the ASME, Journal of Basic
                      Engineering (1960)371-400

[10] Kwakernaak,H.   Linear Optimal Control Systems.
     Sivan,R.   New York: J.Wiley 1972

[11] Lasalle,J.P.   The Stability of Dynamical Systems.
                       Philadelphia: SIAM 1976

[12] Ludyk,G.   Time-Variant Discrete-Time Systems.
                       Braunschweig, Wiesbaden: Vieweg 1981

[13] Ludyk,G.   Über die Stabilität zeitvarianter
                       zeitdiskreter linearer Systeme.
                       Universität Bremen: Berichte Elektro-
                       technik No.1/84 (1984)

[14] Ludyk,G.   Ein Stabilitätskriterium für zeitvari-
                       ante zeitdiskrete lineare Systeme.
                       Regelungstechnik 32(1984)162-163

[15] Miller,K.S.   Linear Difference Equations.
                       New York: Benjamin 1968

[16] Moore,J.B.   Coping with Singular Transition Matrices
     Anderson,B.D.O.   in Estimation and Control Stability
                       Theory.
                       Int.J.Control 31(1980)571-586

[17] Noble,B.   Applied Linear Algebra.
     Daniel,J.W.   Englewood Cliffs: Prentice-Hall 1977

[18] Weiss,L.   Stability of Linear Discrete Time Systems
     Lee,J.-S.   in a Finite Time Interval.
                       Avtomatika i Telemechanika(1971)63-68

[19] Willems,J.L.   Stability Theory of Dynamical Systems.
                       London: Nelson 1970

Index